T0135709

Alexander Sieber

Adaptive Quarklet Schemes

Approximation, Compression, Function Spaces

Logos Verlag Berlin

λογος

Bibliografische Information der Deutschen Nationalbibliothek

Die Deutsche Nationalbibliothek verzeichnet diese Publikation in der
Deutschen Nationalbibliografie; detaillierte bibliografische Daten sind
im Internet über http://dnb.d-nb.de abrufbar.

ISBN 978-3-8325-5196-4

Logos Verlag Berlin GmbH
Georg-Knorr-Str. 4, Geb. 10,
12681 Berlin
Tel.:   +49 (0)30 / 42 85 10 90
Fax: +49 (0)30 / 42 85 10 92
http://www.logos-verlag.de

Dissertation

Adaptive Quarklet Schemes: Approximation, Compression, Function Spaces

Alexander Sieber

2020

Adaptive Quarklet Schemes: Approximation, Compression, Function Spaces

Dissertation
zur
Erlangung des Doktorgrades
der Naturwissenschaften
(Dr. rer. nat.)

dem
Fachbereich Mathematik und Informatik
der Philipps-Universität Marburg
vorgelegt von

M. Sc. Alexander Sieber
aus Gießen

Marburg 2020

Vom Fachbereich Mathematik und Informatik
der Philipps–Universität Marburg (Hochschulkennziffer: 1180)
als Dissertation angenommen am: 10.09.2020

Erstgutachter: Prof. Dr. Stephan Dahlke, Philipps–Universität Marburg

Zweitgutachter: Prof. Dr. Thorsten Raasch, Universität Siegen

Tag der Einreichung: 07.07.2020
Tag der mündlichen Prüfung: 06.10.2020

Acknowledgements

First of all I would like to express my deepest gratitude to Prof. Dr. Stephan Dahlke for his advice and the fruitful collaboration in our research, furthermore for the confidence in my work. Thank you also for the opportunity to participate in conferences and to get in touch with the scientific community. Secondly I am much obliged to Prof. Dr. Thorsten Raasch for the steady cooperation and being my second referee. Furthermore I want to thank Prof. Dr. Peter Oswald for useful hints that helped to improve my work. I also want to thank the whole work group numerics and optimization at Marburg University for the nice and friendly atmosphere, in particular Dr. Philipp Keding with whom I closely worked together on the theory of quarklets and the Wavelet and Multiscale Library. Furthermore I would like to thank my coauthor Dr. Ulrich Friedrich. Finally I want to thank my wife Julia for her support.

Zusammenfassung

Die vorliegende Arbeit thematisiert die numerische Behandlung von Operatorgleichungen mittels adaptiver *Quarklet-Verfahren*. Unter einem Quarklet verstehen wir das Produkt eines Wavelets mit einem stückweisen Polynom, folglich bezeichnen wir eine polynomiell angereicherte Wavelet-Basis als *Quarklet-Frame*. Diese Systeme können auf der reellen Achse und Intervallen beginnend konstruiert und mit Hilfe von Tensorprodukt- und Gebietszerlegungstechniken in höheren Dimensionen verallgemeinert werden. Systeme von Quarklets sind stabil in gewissen Funktionenräumen, besitzen wichtige Kompressionseigenschaften und sind daher grundsätzlich geeignet, um in generischen Diskretisierungs- und Lösungsalgorithmen für Operatorgleichungen Verwendung zu finden. Darüber hinaus geben diese Verfahren als Wavelet-Variante von sogenannten *hp*-Methoden Grund zu der Hoffnung, schnell zu konvergieren.

Problemstellung

Zahlreiche Probleme in Naturwissenschaften und Technik führen auf *partielle Differentialgleichungen*. Beispielsweise kann die Wärmeausbreitung in einem Stab als Lösung einer partiellen Differentialgleichung modelliert werden, die von der Zeit und der eindimensionalen Ortskoordinate abhängt. Die Wärmeleitungsgleichung gehört zur Klasse der parabolischen Differentialgleichungen. Eine weitere Klasse zeitabhängiger Probleme, zum Beispiel die Ausbreitung von Wellen, wird durch hyperbolische Differentialgleichungen modelliert. In dieser Arbeit beschränken wir uns auf den Fall *elliptischer* partieller Differentialgleichungen, welche wiederum stationäre Phänomene, wie die Auslenkung einer Membran oder Platte, beschreiben. Um eindeutige Lösungen zu erhalten, ist dabei zusätzlich die Vorgabe von *Randwerten* nötig. Das Musterbeispiel eines elliptischen Randwertproblems ist die Poisson-Gleichung mit homogenen Dirichlet-Randbedingungen:

$$-\Delta u = g \quad \text{in } D,$$
$$u = 0 \quad \text{auf } \partial D.$$

Hier bezeichnet Δ den Laplace-Operator, g eine stetige Funktion, D eine beschränkte, offene und zusammenhängende Menge mit Rand ∂D und u die unbekannte Lösung. Unter zusätzlichen Annahmen fallen elliptische Randwertprobleme in die allgemeinere Klasse elliptischer Operatorgleichungen in Hilberträumen. Im Gegensatz zu gewöhnlichen Differentialgleichungen, die nur von einer Veränderlichen abhängen, sind partielle Differentialgleichungen erfahrungsgemäß deutlich schwieriger zu lösen. Oft ist es unmöglich, eine geschlossene Form der Lösung anzugeben. Darüber hinaus stellt sich das Konzept klassisch differenzierbarer Funktionen oftmals als zu streng für gewisse Randwertprobleme heraus. Aus diesen Gründen wird ein schwächeres Lösungskonzept behandelt und die Differentialgleichung in eine durch eine Bilinearform $a(\cdot, \cdot)$ induzierte elliptische *Operatorgleichung* in *Sobolev-Räumen* $H_0^m(D)$ überführt:

$$a(u, v) = \langle g, v \rangle_{L_2} \quad \text{für alle } v \in H_0^m(D).$$

Sobolev-Räume sind unendlich-dimensional, folglich kann durch implementierbare Verfahren nur eine *approximative* Lösung berechnet werden.

Etablierte Verfahren

Die vorliegende Arbeit behandelt elliptische partielle Differentialgleichungen in ihrer schwachen Formulierung, wir klammern daher an dieser Stelle Verfahren wie Finite-Differenzen-Verfahren, die Anwendung auf die ursprüngliche Formulierung finden, aus. Zu den lange bekannten Lösungsmethoden für Probleme in ihrer schwachen Formulierung gehört das *Galerkin-Verfahren*. Hier wird das Problem auf einem endlich-dimensionalen Teilraum des eigentlichen Lösungsraums betrachtet. Dieser Teilraum ist mit einer Basis $\{b_i\}_{i=1,\dots,N}$ ausgestattet. Wenn der Ansatzraum dem Lösungsraum in gewissem Sinne nahe kommt, lässt sich die Güte der Näherungslösung durch den Fehler der Bestapproximation im Ansatzraum garantieren:

$$\|u - u_N\|_{H_0^m(D)} \lesssim \inf_{w \in V_N} \|u - w\|_{H_0^m(D)}.$$

Hier bezeichnet V_N den N-dimensionalen Ansatzraum und u_N die berechnete Lösung. Prominente Vertreter des Galerkin-Verfahrens sind die sogenannten *Finite-Elemente-Methoden (FEM)*. Dabei werden lokale Basisfunktionen konstruiert, indem das betrachtete Gebiet unterteilt wird. Bei einer feineren Unterteilung wächst die Dimension des Ansatzraums und die Näherungslösung wird genauer. Der zugrunde-liegende Ansatz ist also *geometrisch*. Die Gitterweite wird üblicherweise mit h bezeichnet, daher rührt die Einordnung der beschriebenen Finite-Elemente-Methoden als h-Methoden. Naheliegende Ansatzfunktionen für Differentialgleichungen zweiter Ordnung sind stückweise lineare Funktionen. Finite-Elemente-Methoden, die an Stelle einer geometrischen Verfeinerung auf der Erhöhung des lokalen Polynomgrads basieren, werden als p-FEM bezeichnet. Auch eine Kombination beider Strategien, sogenannte hp-Methoden, finden Anwendung, bei hp-FEM wurde in Spezialfällen sogar exponentielle Konvergenz beobachtet.

Ein weiterer entscheidender Aspekt ist *Adaptivität*. Uniforme Verfahren, die die Gitterweite oder den Polynomgrad für alle Ansatzfunktionen gleich behandeln, verschwenden wahrscheinlich Freiheitsgrade und damit Ressourcen in Bereichen des Gebiets, in denen die Lösung bereits gut approximiert wird. Adaptive Verfahren sind selbst-steuernd in dem Sinn, dass eine berechnete Lösung auf ihre Genauigkeit hin überprüft und dann in kritischen Regionen gezielt verfeinert wird. Mittlerweile ist bekannt, dass adaptive Verfahren überlegen sind, wenn die Lösung der Operatorgleichung in gewissen Besov-Räumen enthalten ist. Insbesondere zu h-FEM gibt es zahlreiche Lehrbücher und Originalarbeiten, wir verweisen hier auf [17, 54, 76, 89] für einen Überblick. Für hp-FEM sei auf [8–12, 47, 62] verwiesen.

Im Gegensatz zu den geometrischen Finite-Elemente-Methoden verwenden *Wavelet-Verfahren* einen Basis-orientierten Ansatz. Wavelet-Basen sind spezielle Basen für Funktionenräume, unter anderem für Sobolev-Räume. Mit ihrer Hilfe lassen sich die betrachteten Operatorgleichungen also auch diskretisieren, wobei die analog entstehende *Steifigkeitsmatrix* ein Operator zwischen Folgenräumen ist. Spezielle Eigenschaften der Wavelet-Basis induzieren *Komprimierbarkeit* des Operators und theoretisch durchführbare iterative Lösungsalgorithmen werden implementierbar. Bei Wavelet-Verfahren liegt die Adaptivität in der approximativen Anwendung der Steifigkeitsmatrix auf einen Vektor. Dadurch werden einige Probleme, die bei adaptiven FEM auftauchen, komplett umgangen. Wir verweisen auf [18–20, 81]. Üblicherweise werden Wavelets durch dyadische Dilatation und Translation einer einzigen Funktion konstruiert. Daher können sie den h-Verfahren zugerechnet werden. Eine klassische Wavelet-Basis Ψ des Lebesgue-Raums $L_2(\mathbb{R})$ hat die Gestalt

$$\Psi := \left\{ 2^{j/2} \psi(2^j \cdot -k) : j, k \in \mathbb{Z} \right\}.$$

Ein wichtiger Vorteil von Wavelets ist der Umstand, dass man stabile Funktionensysteme in Besov-Räumen allein durch geeignetes Gewichten der Wavelets erhält. Unter zusätzlichen Annahmen lassen sich Besov-Räume $B_q^s(L_q(\mathbb{R}))$ mittels Wavelets charakterisieren durch die äquivalente Norm

$$\|f\|_{B_q^s(L_q(\mathbb{R}))}^q \sim \sum_{j \geq -1} \sum_{k \in \mathbb{Z}} 2^{j(s+\frac{1}{2}-\frac{1}{q})q} |\langle f, \psi_{j,k}\rangle_{L_2(\mathbb{R})}|^q.$$

Quarklet-Verfahren

Ein großer Teil dieser Arbeit wird der naheliegenden Frage gewidmet, ob Wavelet-Versionen von hp-Verfahren zur Behandlung von Operatorgleichungen eingesetzt werden können. Diese Aufgabe lässt sich in einige Bausteine gliedern.

Konstruktion eindimensionaler Quarklet-Systeme

Ein erster Schritt hin zu adaptiven Quarklet-Verfahren ist die Konstruktion stabiler Funktionensysteme. *Quarkonial Decompositions* von Funktionenräumen wurden von H. Triebel in [87] eingeführt. In [30] wurde die polynomielle Anreicherung von Zerlegungen der Eins diskutiert und Stabilität in Besov-Räumen gezeigt. Wichtige Hilfsmittel dafür sind L_q-Stabilitätsabschätzungen und *Bernstein-Ungleichungen* für die angereicherten Funktionen. Darüber hinaus werden gewisse Anforderungen, wie

Jackson-Ungleichungen, an das zugrunde liegende Funktionensystem gestellt. Mit der schrittweisen Betrachtung von angereicherten Funktionensystemen wurde in [30] auch eine Beweistechnik präsentiert, die sich auf den Fall zugrundeliegender Wavelet-Basen übertragen lässt.

In [68] wurden Quarklets als polynomiell angereicherte Wavelets erstmals vorgeschlagen, jedoch der Spezialfall von mit Orthogonalpolynomen angereicherten Haar-Wavelets. In [28] wurden Quarklet-Frames auf der reellen Achse studiert und ihre grundsätzliche Eignung für den Einsatz in adaptiven Verfahren nachgewiesen. In [26] wurden schließlich Quarklet-Frames an Intervalle angepasst, wobei der wesentliche Unterschied zur reellen Achse in der Konstruktion von Randfunktionen liegt. Wir beschreiben kurz die Konstruktion. Es wird von einer zugrunde liegenden, biorthogonalen Spline-Wavelet-Basis ausgegangen. Diese Wavelet-Basis wurde mit Hilfe einer Multiresolution Analysis konstruiert, d.h., die Wavelet sind aus Generatoren, in diesem Fall B-Splines, zusammengesetzt. Dann werden *Quarks* φ_p als Produkt eines monischen Polynoms und dem B-Spline φ definiert:

$$\varphi_p := \left(\frac{\cdot}{\lceil \frac{m}{2} \rceil} \right)^p \varphi, \quad p \in \mathbb{N}_0.$$

Damit wird wiederum ein Quarklet ψ_p eingeführt als

$$\psi_p := \sum_{k \in \mathbb{Z}} b_k \varphi_p (2 \cdot -k),$$

wobei $b = \{b_k\}_{k \in \mathbb{Z}}$ die ursprüngliche Wavelet-Maske bezeichnet. Da die Quarklets als Linearkombinationen von Quarks definiert sind, ist es naheliegend, zuerst die Eigenschaften der Quarks zu studieren. Dazu gehört Verfeinerbarkeit in dem Sinne, dass Vektoren von Quarks verfeinerbar sind. In [31] wurde zudem beschrieben, wie das Quark-Setting in das allgemeinere Multigenerator- und Multiwavelet-Konzept passt. Weitere wichtige Eigenschaften sind, wie oben beschrieben, Stabilität der Einzelskalenfunktionen, Bernstein-Ungleichungen und verschwindende Momente. Damit lassen sich Normäquivalenzen in Besov-Räumen herleiten. Diese Ergebnisse werden in Theorem 4.27 and 4.28 formuliert. Mit Hilfe geeigneter Gewichtung impliziert diese Normäquivalenz bereits die Frame-Eigenschaft in L_2-Sobolev-Räumen:

$$\|f\|_{H^m}^2 \sim \sum_{\lambda \in I} |\langle f, w_\lambda \psi_\lambda \rangle_{H^m}|^2.$$

Quarklet-Frames auf Gebieten

Eine große Klasse elliptischer Operatorgleichungen wird auf Gebieten betrachtet, die sich aus diffeomorph transformierten Einheitskuben zusammensetzen lassen. Wir betrachten hier nur den Fall translatierter Einheitskuben, für allgemeinere Gebiete verweisen wir auf [39]. Zuerst ist dazu der Fall eindimensionaler Quarklet-Frames auf

Kuben zu verallgemeinern. Dafür bietet sich ein Tensorprodukt-Ansatz an, denn die L_2-Sobolev-Räume lassen sich im Zweidimensionalen beschreiben durch

$$H^m([0,1]^2) = H^m([0,1]) \otimes L_2([0,1]) \cap L_2([0,1]) \otimes H^m([0,1]).$$

Mit der folgenden zweistufigen Technik lassen sich auch allgemeinere Frames auf Kuben konstruieren. Zuerst werden Frames für Tensorprodukte von Hilberträumen als Tensorprodukte von Frames gewonnen. Dabei ist die geeignete Angabe von Gewichten für die Frame-Elemente entscheidend. Danach werden Frames, die eine Riesz-Basis enthalten, geeignet umgewichtet, um einen Frame für den Schnitt von Hilberträumen zu erhalten. Schließlich werden Gebiete, wie zum Beispiel das L-Gebiet, das einen wichtigen Testfall für adaptive Verfahren darstellt, nicht-überlappend in Kuben zerlegt. Ausgehend von der Existenz eines Quarklet-Frames auf jedem Kubus wird ein Frame auf dem gesamten Gebiet konstruiert. Dabei wird ähnlich wie im in [13] beschriebenen Wavelet-Fall vorgegangen. Zuerst werden die Kuben mit geeigneten Randbedingungen gewählt. Dann werden Funktionen von benachbarten Kuben geeignet fortgesetzt. Für Differentialgleichungen zweiter Ordnung erweist sich die einfache Spiegelung bereits als geeigneter Fortsetzungsoperator. Das Hauptresultat dieses Kapitels findet sich in Theorem 5.15.

Komprimierbarkeit

Nach der Konstruktion von Frames in Hilberträumen lassen sich diese in generischen Lösungsverfahren für Operatorgleichungen verwenden, sofern sie gewisse Kompressionseigenschaften erfüllen. Das bedeutet, dass die unendliche Steifigkeitsmatrix $\boldsymbol{A}$ sich geeignet durch endlich-dimensionale Matrizen $\boldsymbol{A}_J$ approximieren lassen muss:

$$\sum_{J \in \mathbb{N}} 2^{Js} \|\boldsymbol{A} - \boldsymbol{A}_J\| < \infty, \quad \text{für alle } 0 \le s \le s^*.$$

In [28] wurde die Komprimierbarkeit für den eindimensionalen Laplace-Operator gezeigt, in [26] wurde dieses Resultat auf höhere Dimensionen verallgemeinert. Der Laplace-Operator ist ein wichtiger Spezialfall elliptischer Operatoren, grundsätzlich lassen sich die Kompressionsergebnisse aber auf allgemeinere Operatoren übertragen. Die Komprimierbarkeit beruht wesentlich auf den verschwindenden Momenten der Quarklets und ist entscheidend für die Effizienz adaptiver Quarklet-Verfahren. Um schnellere Konvergenz zu erzielen, wird des Weiteren zwischen *first* und *second compression* unterschieden. Second compression für den Wavelet-Fall wurde zum Beispiel in [34, 80] betrachtet. Bei ausgefeilteren Kompressionsstrategien kann der Umstand, dass die Quarklets eine lokal höhere Regularität besitzen, genutzt werden, um bessere Ergebnisse zu erzielen. Die Hauptergebnisse dieses Kapitels werden in den Theoremen 6.7, 6.15, 6.17 und 6.18 präsentiert.

Approximation von Singularitätenfunktionen

Um exponentielle Konvergenz von Quarklet-Verfahren nachzuweisen, sind natürlich noch weitere Betrachtungen nötig. Eine notwendige Voraussetzung ist die gute direkte Approximation von gewissen Funktionen, die Lösung einer elliptischen Differentialgleichung sind. Es ist bekannt, dass Gebiete mit einspringenden Ecken singuläre Lösungen implizieren. Dafür wird die Funktion x^α mit $\alpha > \frac{1}{2}$ als Modell verwendet. Wir weisen nach, dass sich diese Funktion mit Quarklets approximieren lässt, und zwar so, dass der L_2- bzw. H^1-Fehler exponentiell in Abhängigkeit von den verwendeten Freiheitsgraden fällt. Dazu wird, ähnlich wie in [2], das Intervall $[0,1]$ mit einer geometrischen Partitionierung unterteilt und ein Spline mit variierendem Polynomgrad konstruiert. Dieser Spline wird in Einzelskalenfunktionen entwickelt und dann in Quarklet-Frame-Elemente transformiert. Für diese Transformation betrachten wir zwei verschiedene Methoden, nämlich ein allgemeines Rekonstruktionsprinzip für Multi-Generatoren und Multi-Wavelets sowie ein spezielles Rekonstruktionsprinzip, das eine verbesserte asymptotische Schranke liefert. Wir formulieren die Ergebnisse dieses Kapitels in den Theoremen 7.9, 7.11 und 7.13.

Rekonstruktion von Multi-Generatoren

Vektoren von Quarks bzw. Quarklets passen in das Konzept von Multi-Generatoren bzw. Multi-Wavelets. Dieses allgemeine Konzept findet zum Beispiel Anwendung in der Konstruktion von orthonormalen, interpolierenden Wavelets, das sich mit einzelnen Generatoren nicht realisieren lässt. Wir verwenden Fourier-Techniken, um allgemeine Bedingungen für die Rekonstruktion von Multi-Generatoren herzuleiten und wenden dieses Prinzip auf Multi-Quarks an. Ein allgemeines Resultat findet sich in Theorem 3.2, der Spezialfall der Multi-Quarks in Theorem 4.15.

Aufbau der Arbeit

Teile dieser Arbeit wurden in [26, 27, 31] veröffentlicht. Unser Vorgehen ist wie folgt gegliedert. In Kapitel 1 wiederholen wir die benötigten Grundlagen, dazu gehört die Einführung von Sobolev- und Besov-Räumen in Abschnitt 1.1, partielle Differentialgleichungen und ihre schwache Lösungstheorie in Abschnitt 1.2 und einige Aussagen über Frames in Abschnitt 1.3. Letztere werden insbesondere bei der Fortsetzung von Kubus-Frames auf Gebiete benötigt. In Kapitel 2 betrachten wir einige Aspekte der Wavelettheorie, im Wesentlichen die Konstruktion von biorthogonalen Wavelet-Basen für die Räume $L_2(\mathbb{R})$ in Abschnitt 2.1 und $L_2(0,1)$ in Abschnitt 2.2. Die allgemeine Konstruktion wird anhand zweier Spline-Wavelet-Basen näher erläutert, nämlich die CDF-Basis aus [21] im Fall $L_2(\mathbb{R})$ und die Primbs-Basis aus [66] im Fall $L_2(0,1)$. Der Nachweis von gewissen Rekonstruktionseigenschaften ist dabei ein notwendiger Schritt. Diese Eigenschaften werden in Kapitel 3 auf den Fall von Multi-Wavelets übertragen. In Kapitel 4 betrachten wir Quarklets im Eindimensionalen. Um Wiederholungen zu vermeiden und weil beide Varianten viele Gemeinsamkeiten haben, bemühen wir uns um eine geschlossene Behandlung von Quarks auf der reellen Achse bzw. dem Intervall in Abschnitt 4.1. In Abschnitt 4.2 wenden wir die Ergebnisse aus Kapitel 3 an und leiten Rekonstruktionsformeln für Quarklets auf der reellen Achse her. Für die Konstruktion der Randquarklets ist zusätzliche Arbeit notwendig, dies geschieht in Abschnitt 4.3. Der Rest von Kapitel 4 behandelt die Stabilität von Quarklets in Funktionenräumen. In Abschnitt 4.4 bzw. 4.7 wiederholen wir die bekannten Ergebnisse für die Räume L_2 und H^s. Wir verallgemeinern diese Ergebnisse für Besov-Räume in Abschnitt 4.5 und 4.6. Kapitel 5 ist Quarklet-Frames für Sobolev-Räume auf Gebieten gewidmet, nämlich Einheitskuben in Abschnitt 5.1 und zusammengesetzten Gebieten in Abschnitt 5.2. Damit lassen sich adaptive Quarklet-Verfahren entwerfen, die wir in Kapitel 6 beschreiben. Nach einigen allgemeinen Aspekten solcher Verfahren in Abschnitt 6.1 zeigen wir die Komprimierbarkeit für den Laplace-Operator mit Hilfe von Quarklet-Frames in Abschnitt 6.2 und 6.3. Dabei betrachten wir insbesondere Second-Compression-Ansätze. Schließlich diskutieren wir die direkte Approximation von singulären Lösungen mittels Quarklets in Kapitel 7. In Abschnitt 7.1 weisen wir exponentielles Abfallverhalten des L_2-Approximationsfehlers nach, in Abschnitt 7.2 studieren wir den H^1-Fehler. Wir übertragen diese Ergebnisse auf den höherdimensionalen Fall in Abschnitt 7.3. Die Arbeit schließt mit einigen numerischen Experimenten zu den besprochenen Aspekten adaptiver Quarklet-Verfahren in Kapitel 8. Als üblicher Testfall wird in Abschnitt 8.1 die Poisson-Gleichung auf zweidimensionalen Gebieten mit einspringenden Ecken numerisch gelöst. Des weiteren betrachten wir numerische Experimente zu Second Compression in Abschnitt 8.2 und direkter Approximation in Abschnitt 8.3.

Contents

Introduction

This thesis is dedicated to the numerical treatment of operator equations using adaptive quarklet methods. By a *quarklet* we understand the product of a wavelet and a piecewise polynomial, consequently we call a polynomially enriched wavelet basis *quarklet frame*. These function systems can initially be constructed both on the real line and the unit interval and, having done this, be generalised to higher dimensions by using tensor product techniques and domain decompositions. Quarklet systems are stable in function spaces, furthermore they fulfil certain compressibility properties and hence are convenient to be utilised in generic frame methods for the treatment of operator equations. Furthermore they represent a wavelet version of *hp*-methods, therefore there is strong hope that they converge quite fast.

Scope of Problems

Numerous problems in science and technology can be described with the help of *partial differential equations*. For example, the heat distribution in some material can be described as the solution of a partial differential equation, which is dependent on the time and the space coordinates. The heat equation belongs to the class of parabolic partial differential equations. A further class of time-dependent equations, e.g., the wave equation, are the hyperbolic differential equations. However, we restrict our discussion to the case of *elliptic* partial differential equations, which describe stationary phenomena, such as the deflection of a membrane or the bending of a board. To ensure unique solutions, one additionally has to incorporate boundary values. The most prominent example of an elliptic boundary value problem is the Poisson equation with homogeneous Dirichlet boundary conditions:

$$-\Delta u = g \quad \text{in } D,$$
$$u = 0 \quad \text{on } \partial D.$$

Here, Δ denotes the Laplacian, g a continuous function, $D \subset \mathbb{R}^d$ a bounded, open, and connected set with boundary ∂D and u the unknown solution. Under certain additional assumptions elliptic boundary value problems belong to the class of elliptic operator equations in Hilbert spaces. In contrast to ordinary differential equations which just depend on one variable, partial differential equations are rather hard to solve analytically. Very often there does not exist a closed form of the solution. Furthermore, the concept of classical differentiable functions turns out to be too restrictive for certain boundary value problems. To overcome this obstacle, one treats a weaker concept of solutions. The boundary value problem is transformed into an operator equation on a *Sobolev space* $H_0^m(D)$ which is induced by a bilinear form $a(\cdot, \cdot)$:

$$a(u, v) = \langle g, v \rangle_{L_2} \quad \text{for all } v \in H_0^m(D).$$

Since the Sobolev spaces are infinite-dimensional generally, one can only gain an approximate solution by implementable methods.

Established Methods

Since this thesis is concerned with elliptic boundary value problems in their weak formulation, we exclude solution methods like finite difference methods that treat the classical formulation. The *Galerkin scheme* is a well-established method for the solution of elliptic boundary value problems. There the problem is considered on a finite-dimensional subspace of the solution space. This subspace is equipped with a basis $\{b_i\}_{i=1,\dots,N}$. With a bijection between $\mathbb{R}^d$ and this space, one can switch to a system of linear equations with coefficient matrix $(a(b_j, b_i))_{i=1,\dots,N, j=1,\dots,N}$. If the ansatz space in a certain sense is close to the solution space, then one can guarantee that the approximate solution u_N is close to the exact one:

$$\|u - u_N\|_{H_0^m(D)} \lesssim \inf_{w \in V_N} \|u - w\|_{H_0^m(D)}.$$

Here, V_N denotes the N-dimensional subspace and u_N the solution computed on this space. Prominent representatives of Galerkin schemes are the *finite element methods* (FEM). There local basis functions are constructed associated with a decomposition of the domain. A finer decomposition leads to a higher dimension of the ansatz space and a more precise solution. Hence finite element methods provide a geometric concept. The width of the mesh is usually denoted by h, therefore we denominate methods that are based on a solution space decomposition with respect to the space coordinates as h-methods. For second order elliptic partial differential equations, piecewise linear ansatz functions are an obvious choice. Finite element methods that rely on increasing the polynomial degree of the ansatz functions are known as p-FEM. Even a combination of both versions is possible. For special cases of so-called hp-FEM exponential convergence has been achieved.

Furthermore, *adaptivity* plays a key role. In uniform schemes the refinement strategy is equally performed for all ansatz functions. Therefore, degrees of freedom and hence computing capacity are wasted in regions of the domain where the solution is already well approximated. Adaptive schemes are self-controlled in the sense that the exactness of a calculated solution is estimated and the refinement is performed only in regions where the error is probably large. Meanwhile it is well-known that adaptive schemes are superior if the solution of the partial differential equation is contained in certain Besov spaces. In particular for h-FEM, there is a huge amount of literature, we refer to [17, 54, 76, 89] for an overview. For hp-FEM we refer to [8–12, 47, 62].

In contrast to the geometric concept of finite element methods, *wavelet methods* represent a basis orientated concept. Wavelet bases are certain bases for function spaces, in particular for Sobolev spaces. They provide strong analytical properties. It is possible to construct wavelets with arbitrarily high but finite regularity and still compact support. Furthermore they possess a certain amount of vanishing moments which lead to sparse representations of smooth functions. Wavelets have successfully been employed in techniques of signal analysis, but they can also be utilised for the discretisation of operator equations. In this case the resulting stiffness matrix is biinfinite and can be interpreted as an operator between sequence spaces. Certain properties of the wavelet basis and of the operator induce *compressibility* of the operator, therefore one can implement inexact versions of the theoretically feasible iterative solution methods. In the case of wavelet methods, the adaptivity reduces to approximately applying the stiffness matrix to some vector. Therefore, adaptive wavelet schemes are not faced with many typical problems that arise with adaptive finite element methods. We refer to [18–20, 81]. Usually wavelets are constructed by dyadic dilation and translation of a single function. Hence they belong to the class of h-methods. A classical wavelet basis Ψ of $L_2(\mathbb{R})$ has the form

$$\Psi := \left\{ 2^{j/2} \psi(2^j \cdot -k) : j, k \in \mathbb{Z} \right\}.$$

With certain additional assumptions, the classical wavelet characterisation of a Besov space reads as

$$\|f\|_{B_q^s(L_q(\mathbb{R}))}^q \sim \sum_{j \geq -1} \sum_{k \in \mathbb{Z}} 2^{j(s+\frac{1}{2}-\frac{1}{q})q} |\langle f, \psi_{j,k} \rangle_{L_2(\mathbb{R})}|^q.$$

Quarklet Schemes

A large part of this thesis will be concerned with the obvious question if it is possible to design wavelet versions of hp-methods for the solution of elliptic operator equations. This task can be decomposed into several smaller steps.

Construction of Univariate Quarklet Systems

A first step to establish adaptive quarklet methods is to construct stable function systems in the univariate setting. *Quarkonial decompositions* of function spaces have been invented by H. Triebel in [87]. In [30] the polynomial enrichment of partitions of unity has been discussed and their stability in Besov spaces was shown. Important tools are L_q stability and *Bernstein inequalities* of the enriched functions. Further assumptions on the underlying system were formulated, e.g., *Jackson inequalities*. In loc. cit. there also was presented a proof technique of successively considering higher enriched systems which can be transferred to the case of an underlying wavelet basis.

In [68] quarklets were primarily mentioned as Haar wavelets which are enriched by orthogonal polynomials. In [28] quarklet frames on the real line based on the CDF wavelet basis have been discussed. In particular their compressibility and applicability for adaptive frame schemes were proven. Finally, in [26] quarklet frames on the interval have been constructed, where the major difference to the shift-invariant case lies in the appropriate construction of boundary functions. We briefly describe the construction of univariate quarklets. It is based on an underlying biorthogonal spline wavelet basis. This wavelet basis has been constructed with the help of a multiresolution analysis, that means the wavelets are linear compositions of generators. In our particular case those generators are B-splines. Then, a *quark* φ_p is defined as the product of a monic polynomial and a B-spline φ:

$$\varphi_p := \left(\frac{\cdot}{\lceil \frac{m}{2} \rceil} \right)^p \varphi, \quad p \in \mathbb{N}_0.$$

Then, the quarklet ψ_p is defined by

$$\psi_p := \sum_{k \in \mathbb{Z}} b_k \varphi_p(2 \cdot - k),$$

where $b = \{b_k\}_{k \in \mathbb{Z}}$ denotes the original wavelet mask. Since the quarklets are linear combinations of quarks, it suffices to primarily study the properties of the quarks. One of these properties is refinability in the sense that vectors of quarks are refinable. In [31] it has been shown how these multiquarks fit in the framework of more general multigenerators and multiwavelets. Further important properties are, as mentioned above, stability properties of the single-scale functions, Bernstein inequalities and vanishing moments. These properties allow to deduce equivalent norms for Besov spaces. These results are stated in Theorem 4.27 and 4.28. With suitable weights the frame property in L_2-Sobolev spaces immediately follows:

$$\|f\|_{H^m}^2 \sim \sum_{\lambda \in I} |\langle f, w_\lambda \psi_\lambda \rangle_{H^m}|^2.$$

Quarklet Frames on Domains

A large class of elliptic operator equations is considered on domains which are decomposable into diffeomorphic transformed unit cubes. We restrict our discussion to the case of translated cubes, for more general domains we refer to [39] . In a first step of generalisation one has to construct quarklet frames on cubes. A tensor product approach is favourable since the Sobolev spaces in two dimensions provide the following structure which can be generalised to higher dimensions:

$$H^m([0,1]^2) = H^m([0,1]) \otimes L_2([0,1]) \cap L_2([0,1]) \otimes H^m([0,1]).$$

By performing the following two steps one can even construct more general tensor product frames on cubes. At first, tensor products of frames are frames for tensor products of Hilbert spaces if the weights are appropriately chosen. Subsequently, rescaled frames which contain a Riesz basis are frames for intersections of Hilbert spaces. Afterwards, domains like the L-shaped domain, which represents a prominent test case for adaptive algorithms, are decomposed into cubes in a non-overlapping way. Based upon a quarklet frame on each cube a quarklet frame on the whole domain is constructed. We proceed similar to the wavelet case described in [13]. Initially, one has to choose boundary conditions for the cubes. Then, quarklets defined on a cube are suitably extended to the neighbouring cubes. For second order differential equations the simple reflection turns out to be a suitable extension operator. The main result of this chapter is presented in Theorem 5.15.

Compressibility

Having constructed frames of quarklet type, they are ready to be utilised in generic frame methods if they possess certain compression properties. That means, the biinfinite stiffness matrix A has to be well approximated by finite matrices A_J:

$$\sum_{J \in \mathbb{N}} 2^{Js} \|A - A_J\| < \infty \quad \text{for all } 0 \leq s \leq s^*.$$

In [28] compressibility for the univariate Laplacian has been shown, in [26] this result has been generalised to higher dimensions. The Laplacian is an important special case, nevertheless the compression results can be transferred to other elliptic operators. The compressibility heavily relies on the vanishing moments property of the quarklets and affects the performance of adaptive quarklet schemes. To achieve faster convergence one can distinguish between *first* and *second compression*. Second compression in the case of wavelets has been studied in [34, 80]. One can use the fact that the quarklets possess a higher local regularity in the design of more involved compression strategies to gain better results. The univariate main result of this Section is presented in Theorem 6.7. The different strategies and results in the multivariate case are described in Theorems 6.15, 6.17, 6.18.

Approximation of Singular Solutions

It is of course a long way to go to achieve provable exponential convergence of adaptive quarklet schemes, though a necessary condition is that typical solutions to partial differential equations can be well approximated in terms of quarklets. It is well known that domains with re-entrant corners, such as the L-shaped domain, induce singular solutions. The univariate function x^α with $\alpha > \frac{1}{2}$ serves as a model for such solutions. We show that this particular function can be approximated in terms of quarklets such that the error decreases exponentially with respect to degrees of freedom. To prove this, similar to [2], a certain spline with varying polynomial degree is constructed on a partition of the interval $[0, 1]$ that geometrically becomes finer advancing to the left boundary. In contrast to that result, the spline described here is $m - 2$ times continuously differentiable. The construction can be found in Theorem 7.1. This particular spline can be developed into single-scale quarks and, with the help of reconstruction properties, transformed into quarklet frame elements. To perform these transformations, we consider two different methods, firstly a general reconstruction principle that fits in the framework of general multiwavelets, secondly an adapted reconstruction principle that even provides a better asymptotic error decay. The results for the cases $L_2(I)$, $H^1(I)$ and $H^1(I^2)$ are presented in the Theorems 7.9, 7.11, 7.13, respectively.

Reconstruction of Multigenerators

Vectors of quarks or quarklets fit in the framework of multigenerators and multiwavelets, respectively. This general concept is, e.g., applied to the construction of orthonormal interpolating wavelets which can not be realised with a single generator. We use Fourier techniques to derive a general criterion for the reconstruction of multigenerators in Theorem 3.2 and apply this result to the special case of multiquarks in Theorem 4.15.

Layout

Parts of this thesis have been published in [26, 27, 31]. We proceed in the following way. In Chapter 1 we treat the necessary foundations. In Section 1.1 we introduce function spaces, in particular Sobolev and Besov spaces. In Section 1.2 we recall some basic facts about elliptic partial differential equations and their weak formulation. Section 1.3 is dedicated to frames. In Chapter 2 we recall the construction of biorthogonal wavelet bases. Firstly we consider the classical construction of a wavelet basis of $L_2(\mathbb{R})$, secondly we treat the construction of a wavelet basis on the interval. Two particular spline wavelet bases are of our interest, namely the CDF wavelet basis from [21] and the basis constructed by M. Primbs in [66]. There reconstruction properties appear as a by-product. In Chapter 3 we derive a general criterion for multigenerators to fulfil certain reconstruction properties. In Chapter 4 we introduce univariate quarklets. In Section 4.1 we recall the construction and properties of quark generators on the real line and the interval, respectively. In Section 4.2 we apply the results from Chapter 3 to derive reconstruction properties of the shift-invariant quarklets. In Section 4.3 we adapt quarklets to the interval. This is done by the appropriate construction of boundary functions. The remainder of this chapter is dedicated to the stability of quarklet systems in function spaces, in particular in Section 4.5 and 4.6 we derive equivalent norms for Besov spaces. The known special cases of L_2 and L_2-Sobolev spaces are recalled in Section 4.4 and 4.7, respectively. Having shown the frame property of quarklet systems, it is possible to design adaptive quarklet schemes. This is done in Chapter 6. In Sections 6.2, 6.3 the crucial compression properties of quarklet frames are treated, where we pay attention to second compression strategies. Finally, in Chapter 7 we study approximation properties of the quarklets in L_2 and H^1 norms, respectively. Using the tensor product ansatz from Section 5.1, it suffices to consider univariate quarklet approximation for certain singular solutions to elliptic boundary value problems. The thesis closes with several numerical experiments in Chapter 8. As a canonical example we treat the Poisson equation on two-dimensional domains with re-entrant corners numerically. Furthermore, second compression and quarklet approximation are tested.

Chapter 1

Preliminaries

In this chapter we present definitions and results that are needed in the course of this thesis. In the first section we discuss the relevant function spaces for the theory of adaptive quarklet schemes, in Section 1.2 we introduce the class of operator equations which we are going to treat. Finally, in Section 1.3 we recall the concept of frames.

At first, we fix some notation. As usual, let $\mathbb{N}$, $\mathbb{Z}$, $\mathbb{R}$, $\mathbb{C}$ denote the set of all natural, integer, real and complex numbers. Furthermore, we define $S_1 := \{z \in \mathbb{C} : |z| = 1\}$ as the unit sphere in the complex plane which can be identified with the torus $T = \mathbb{R}/_{2\pi\mathbb{Z}}$. The frequently used relation $a \lesssim b$ is understood as $a \leq Cb$ with a constant $C > 0$ that does not depend on a and b. The relation $a \sim b$ is fulfilled if $a \lesssim b$ and $b \lesssim a$.

1.1 Function Spaces

In this section we introduce the relevant function spaces for solving elliptic partial differential equations. In particular we consider Lebesgue spaces, Sobolev spaces and Besov spaces. We always assume $D \subset \mathbb{R}^d$ to be a domain. Unless otherwise stated, the findings in this section stem from [1, 45, 54, 84–86, 88, 90].

Definition 1.1. Let $p > 0$. Then, the spaces $\mathcal{L}_p(D)$ are defined by

$$\mathcal{L}_p(D) := \{f : D \to \mathbb{R} \text{ measurable}, \|f\|_{\mathcal{L}_p(D)} < \infty\}, \tag{1.1.1}$$

$$\|f\|_{\mathcal{L}_p(D)} := \begin{cases} \left(\int_D |f(x)|^p \, dx\right)^{1/p}, & 0 < p < \infty, \\ \operatorname*{ess\,sup}_{x \in D} |f(x)|, & p = \infty. \end{cases} \tag{1.1.2}$$

Considering equivalence classes of functions, we obtain the *Lebesgue spaces* $L_p(D)$ as the quotient spaces

$$L_p(D) := \mathcal{L}_p(D)/_N, \tag{1.1.3}$$

$$\|[f]\|_{L_p(D)} := \|f\|_{\mathcal{L}_p(D)}, \tag{1.1.4}$$

where $N := \{f : f \in \mathcal{L}_p(D), \|f\|_{\mathcal{L}_p(D)} = 0\}$ denotes the set of functions that are equal to zero almost everywhere.

9

Note that for $p \geq 1$ the L_p spaces are Banach spaces, where for $0 < p < 1$ they are quasi-Banach spaces since the triangle inequality only holds up to a constant. The space $L_2(D)$ provides further structure since it is a Hilbert space with inner product

$$\langle f, g \rangle_{L_2(D)} := \int_D f(x)\overline{g(x)} \, dx. \tag{1.1.5}$$

For the construction of wavelets we consider the spaces $\mathscr{L}_p$ which contain functions with a certain summation property:

$$\mathscr{L}_p := \{ f : \sum_{k \in \mathbb{Z}^d} |f(\cdot - k)| \in L_p([0,1]^d) \}. \tag{1.1.6}$$

The space of *locally integrable functions* $L_{1,loc}(D)$ is defined by

$$L_{1,loc}(D) := \{ f : f \in L_1(K) \text{ for all } K \subset\subset D \}. \tag{1.1.7}$$

To define Sobolev spaces, we introduce further notation. We call $\alpha \in \mathbb{N}_0^d$ a *multiindex* with absolute value $|\alpha| := \sum_{i=1}^d \alpha_i$. Then, the α-th derivative of a differentiable function $f \in C^{|\alpha|}(D)$ has the form $D^\alpha f := \frac{\partial^{|\alpha|}}{\partial x_1^{\alpha_1} \cdots \partial x_d^{\alpha_d}} f$. The *support* of a function f is defined as the closure of the set of all points where f differs from zero:

$$\text{supp} f := \overline{\{ x : x \in D, f(x) \neq 0 \}}. \tag{1.1.8}$$

The space of *test functions* $C_0^\infty(D)$ is defined as the collection of all infinitely often differentiable functions with compact support. With these preparations, we are able to state the concept of *weak derivatives*, which is the classical approach to Sobolev spaces. A function $f_\alpha \in L_{1,loc}(D)$ is called the *α-th weak derivative* of $f \in L_{1,loc}(D)$ if it holds

$$\int_D f_\alpha(x)\varphi(x) \, dx = (-1)^{|\alpha|} \int_D f(x) D^\alpha \varphi(x) \, dx \quad \text{for all } \varphi \in C_0^\infty(D). \tag{1.1.9}$$

Note that the weak derivative is uniquely determined almost everywhere if it exists. For classically differentiable functions, the weak and classical derivative coincide. In the following we denote both classical and weak derivatives with D^α.

Definition 1.2. Let $m \in \mathbb{N}_0$ and $p \geq 1$. The *classical Sobolev spaces* $W_p^m(D)$ are defined as the collection of all functions which have weak derivatives up to order m in $L_p(D)$:

$$W_p^m(D) := \{ f : D^\alpha f \in L_p(D) \text{ for all } 0 \leq |\alpha| \leq m \}, \tag{1.1.10}$$

$$\|f\|_{W_p^m(D)} := \begin{cases} \left(\sum_{|\alpha| \leq m} \|D^\alpha f\|_{L_p(D)}^p \right)^{1/p}, & 1 \leq p < \infty, \\ \sup_{|\alpha| \leq m} \|D^\alpha f\|_{L_p(D)}, & p = \infty. \end{cases} \tag{1.1.11}$$

Under certain conditions, an equivalent norm for Sobolev spaces is given by

$$\|f\|^*_{W_p^m(D)} := \|f\|_{L_p(D)} + |f|_{W_p^m(D)}, \tag{1.1.12}$$

$$|f|_{W_p^m(D)} := \left(\sum_{|\alpha|=m} \|D^\alpha f\|^p_{L_p(D)} \right)^{1/p}, \tag{1.1.13}$$

see [1, Cor. 4.16]. It is also possible to define Sobolev spaces with positive, real-valued regularity.

Definition 1.3. Let $s \in \mathbb{R}_+ \setminus \mathbb{N}$, $m := \lfloor s \rfloor$, $\sigma := s - m$ and $p \geq 1$. The *Sobolev spaces* $W_p^s(D)$ are defined by

$$W_p^s(D) := \left\{ f : f \in W_p^m(D), \|f\|_{W_p^s(D)} < \infty \right\}, \tag{1.1.14}$$

$$\|f\|_{W_p^s(D)} := \left(\|f\|_{W_p^m(D)} + \sum_{\alpha=m} |D^\alpha f|^p_{\sigma,p,D} \right)^{1/p}, \tag{1.1.15}$$

where

$$|f|^p_{\sigma,p,D} := \int_D \int_D \frac{|f(x) - f(y)|^p}{|x - y|^{d+\sigma p}} \, \mathrm{d}x \, \mathrm{d}y. \tag{1.1.16}$$

The Sobolev spaces are Banach spaces, similar to the case of Lebesgue spaces the case $p = 2$ plays a distinguished role. The spaces $W_2^s(D)$ are Hilbert spaces, this motivates the notation $H^s(D) := W_2^s(D)$. In the case of integer smoothness the $H^m(D)$ scalar product has the form

$$\langle f, g \rangle_{H^m(D)} = \sum_{|\alpha| \leq m} \langle D^\alpha f, D^\alpha g \rangle_{L_2(D)}. \tag{1.1.17}$$

Otherwise, with m and σ as in Definition 1.3, we have

$$\begin{aligned}
\langle f, g \rangle_{H^s(D)} = &\langle f, g \rangle_{H^m(D)} \\
&+ \sum_{|\alpha| \leq m} \int_D \int_D \frac{|D^\alpha f(x) - D^\alpha f(y)||D^\alpha g(x) - D^\alpha g(y)|}{|x - y|^{d+2\sigma}} \, \mathrm{d}x \, \mathrm{d}y.
\end{aligned} \tag{1.1.18}$$

Another common approach to define Sobolev spaces is the Fourier transform. For $f \in L_1(\mathbb{R}^d)$, $\widehat{f}$ is defined by

$$\widehat{f}(\xi) := \int_{\mathbb{R}^d} f(x) e^{-ix\xi} \, \mathrm{d}x, \quad \xi \in \mathbb{R}^d. \tag{1.1.19}$$

Note that, with the help of Schwartz functions, the Fourier transform can be extended to the space $L_2(\mathbb{R}^d)$. With this at hand, one defines

$$\langle f, g \rangle^\wedge_{H^s(\mathbb{R}^d)} := \int_{\mathbb{R}^d} (1 + |x|^2)^s \widehat{f}(x) \overline{\widehat{g}(x)} \, \mathrm{d}x. \tag{1.1.20}$$

11

Then, $\| \cdot \|_{H^s(\mathbb{R}^d)}^{\wedge} := (\langle f, g \rangle_{H^s(\mathbb{R}^d)}^{\wedge})^{1/2}$ provides an equivalent norm, cf. [54, Lem. 6.41]. When it comes to the problem of solving elliptic PDEs, one has to consider Sobolev spaces with incorporated boundary conditions. There are several ways to define these spaces, one possibility is to define the Sobolev space with zero boundary conditions $H_0^m(D)$ as the closure of $C_0^\infty(D)$ with respect to the $H^m(D)$ norm. Let $\Gamma \subset \partial D$ be a subset of the boundary of D. Then, the Sobolev space with incorporated zero boundary conditions on Γ is defined as

$$H_{0,\Gamma}^m(D) := \overline{\{f : f \in C^\infty(D) \cap H^m(D) : f = 0 \text{ in a neighbourhood of } \Gamma\}}^{\|\cdot\|_{H^m(D)}}.$$
(1.1.21)

In the case $\Gamma = \partial D$ we use the shorthand notation $H_0^m(D) := H_{0,\partial D}^m(D)$. Furthermore, with $H^{-m}(D)$, we denote the dual space of $H_0^m(D)$. Let us specify the class of domains under consideration. Since the regularity of the unknown solution of a partial differential equations depends on the smoothness of the boundary of the domain, we restrict our discussion to Lipschitz domains.

Definition 1.4. [45, 54] Let either $k \in \mathbb{N} \setminus \{0\}$, $\alpha \in [0, 1]$ or $k = 0$, $\alpha = 1$. The domain $D \subset \mathbb{R}^d$ is called $C^{k,\alpha}$ domain, if for each $x \in \Gamma$ there exists a neighbourhood $U \subset \mathbb{R}^d$ and a bijective mapping $\varphi : U \to B_1(0)$ such that it holds

$$\varphi \in C^{k,\alpha}(\overline{U}), \quad \varphi^{-1} \in C^{k,\alpha}(\overline{U}),$$
$$\varphi(U \cap \Gamma) = \{\xi \in B_1(0) : \xi_d = 0\},$$
$$\varphi(U \cap D) = \{\xi \in B_1(0) : \xi_d > 0\},$$
$$\varphi(U \cap (\mathbb{R}^d \setminus D)) = \{\xi \in B_1(0) : \xi_d < 0\}.$$

Domains of the type $C^{0,1}$ are called *Lipschitz domains*.

Now we turn to Besov spaces. We consider the discrete difference operator Δ_h^m for measurable functions, which is defined by

$$\Delta_h f := \Delta_h^1 f := f(\cdot + h) - f(\cdot),$$
$$\Delta_h^m f := \Delta_h \left(\Delta_h^{m-1} f\right), \quad m \geq 2.$$
(1.1.22)

The iterated difference Δ_h^m has the closed form

$$\Delta_h^m f := \left(\prod_{i=0}^m \chi_D(\cdot + ih)\right) \sum_{j=0}^m \binom{m}{j} (-1)^{m-j} f(\cdot + jh), \quad h \in \mathbb{R}^d, m \in \mathbb{N}_0.$$
(1.1.23)

Discrete differences allow us to measure smoothness in a more local sense than weak derivatives which have to exist on the whole domain. The *modulus of smoothness* $\omega_m(f, t)_{L_p(D)}$ of a function f is defined by

$$\omega_m(f, t)_{L_p(D)} := \sup_{|h| \leq t} \|\Delta_h^m f\|_{L_p(D)}, \quad t > 0, m \in \mathbb{N}_0.$$
(1.1.24)

We collect some properties of general moduli of smoothness that are needed later on, cf. [43, Ch. 2.7].

Lemma 1.5. *For $m \in \mathbb{N}$, $f, g \in L_p(D)$ it holds*

$$\omega_m(f+g,t)_{L_p(D)} \leq \begin{cases} \omega_m(f,t)_{L_p(D)} + \omega_m(g,t)_{L_p(D)}, & p \geq 1, \\ \left(\omega_m(f,t)_{L_p(D)} + \omega_m(g,t)_{L_p(D)}\right)^{1/p}, & p < 1. \end{cases} \quad (1.1.25)$$

$$\omega_m(f,t)_{L_p(D)} \leq \begin{cases} 2^m \|f\|_{L_p(D)}, & p \geq 1, \\ 2^{m/p} \|f\|_{L_p(D)}, & p < 1. \end{cases} \quad (1.1.26)$$

With these preparations we are able to introduce Besov spaces.

Definition 1.6. Let $0 < p < \infty$, $0 < q < \infty$, $s > 0$ and $m := \lfloor s \rfloor + 1$. The *Besov spaces* $B_q^s(L_p(D))$ are defined by

$$B_q^s(L_p(D)) := \{f : f \in L_p(D), \|f\|_{B_q^s(L_p(D))} < \infty\}, \quad (1.1.27)$$

$$\|f\|_{B_q^s(L_p(D))} := \|f\|_{L_p(D)} + \left(\int_0^\infty \left(t^{-s}\omega_m(f,t)_{L_p(D)}\right)^q \frac{dt}{t}\right)^{1/q}. \quad (1.1.28)$$

Note that for $p, q \geq 1$ the Besov spaces are Banach spaces. Otherwise, the expression $\|\cdot\|_{B_q^s(L_p(D))}$ is just a quasi-norm and the Besov spaces are quasi-Banach spaces. The following result from [85, Sec. 2.5.12] gives a useful characterisation of Besov spaces by discrete norms.

Theorem 1.7. *The expression*

$$\|f\|^*_{B_q^s(L_p(D))} := \|f\|_{L_p(D)} + \left(\sum_{j=0}^\infty 2^{jsq}\omega_m(f,2^{-j})^q_{L_p(D)}\right)^{1/q} \quad (1.1.29)$$

is an equivalent norm on $B_q^s(L_p(D))$.

For smoothness parameters $s \notin \mathbb{N}$ and $p = q$ Besov and Sobolev spaces coincide, i.e., $W_p^s(D) = B_p^s(L_p(D))$, furthermore for all s it holds $H^s(D) = B_2^s(L_2(D))$. Figure 1.1 shows a so-called *DeVore-Triebel diagram*. It depicts two scales of Besov spaces, namely the uniform scale of the spaces $H^s(D)$ and the *adaptivity scale* of spaces $B_\tau^s(L_\tau(D))$ for parameters

$$\frac{1}{\tau} = \frac{s}{d} + \frac{1}{2}. \quad (1.1.30)$$

These scales determine the convergence rate of numerical solution methods to elliptic PDEs as follows. If the unknown solution belongs to some Sobolev space $H^s(D)$, $s > 0$, the error of a uniform scheme decreases in the order of $N^{-s/d}$. On the other hand, if the unknown solution belongs to some Besov space $B_\tau^\sigma(L_\tau(D))$, the error of an adaptive scheme decreases in the order of $N^{-\sigma/d}$, where N denotes the number of degrees of freedom. That means, for solutions with significant higher Besov regularity, adaptivity pays off.

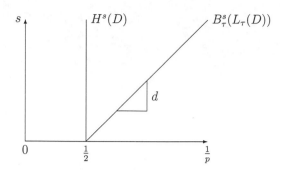

Figure 1.1: Function spaces in a DeVore-Triebel diagram.

1.2 Elliptic Partial Differential Equations

In this section we specify which class of problems is treated in the course of this thesis. We give a brief introduction to the classical setting of elliptic boundary value problems such as (1.2.3). Then we recall the development of a weak formulation (1.2.10) of elliptic BVPs and conditions for the existence and uniqueness of weak solutions. In particular we consider the Laplace operator as the most prominent example of second order elliptic differential operators. For a detailed analysis we refer to [49, 54]. We start with the basic definitions of partial differential equations.

Definition 1.8. Let $D \subset \mathbb{R}^d$ be a domain, $m \in \mathbb{N}$, $u \in C^{2m}(D)$, $\alpha \in \mathbb{N}^d$, $a_\alpha \in L_\infty(D)$. The operator L, given by

$$Lu(x) := \sum_{|\alpha| \leq 2m} a_\alpha D^\alpha u(x), \quad x \in D, \tag{1.2.1}$$

is called *partial differential operator of order* $2m$. The operator L is called *elliptic in* $x \in D$, if for all $\xi \in \mathbb{R}^d$ it holds

$$\sum_{|\alpha|=2m} a_\alpha(x)\xi^\alpha \geq c(x)|\xi|^{2m}. \tag{1.2.2}$$

L is called *uniformly elliptic in* D, if it holds $\inf_{x \in D} c(x) > 0$.

These definitions allow us to formulate the starting point for the considerations in this thesis. A *partial differential equation* (PDE) is given by

$$Lu = g \quad \text{in } D.$$

Similar to the case of ordinary partial differential equations, the solution of a partial differential equation is not unique. For uniqueness one needs further restrictions such as initial or boundary values. In the case of elliptic PDEs, only boundary values are admissible. In the sequel we always assume D to be a bounded domain. A *boundary value problem* (BVP) is given by

$$
\begin{aligned}
Lu &= g \quad \text{in } D, \\
B_j u &= \varphi_j \quad \text{on } \partial D, \; j = 1, \dots, m,
\end{aligned}
\tag{1.2.3}
$$

where B_j are called *boundary differential operators* of order $2m - 1$. In general they are given by

$$
B_j u := \sum_{|\alpha| \le m_j} b_{j_\alpha} D^\alpha u, \quad 0 \le m_j \le 2m - 1.
$$

An important special case are *Dirichlet boundary conditions*:

$$
B_j = \left(\frac{\partial}{\partial n} \right)^{j-1}.
$$

Here n denotes the outer normal vector. A function $u \in C^{2m}(\overline{D})$ fulfilling (1.2.3) is called *classical solution*. The most important example of a second order elliptic BVP is the *Poisson equation* with homogeneous Dirichlet boundary conditions:

$$
\begin{aligned}
-\Delta u &= g \quad \text{in } D, \\
u &= 0 \quad \text{on } \partial D,
\end{aligned}
\tag{1.2.4}
$$

where $\Delta := \sum_{j=1}^{d} \frac{\partial^2}{\partial x_j^2}$ denotes the Laplacian. The approach of classical solutions often turns out to be too restrictive. To overcome this obstacle the approach of *weak solutions* has been developed. In the following we deduce the *weak formulation* of an elliptic BVP with homogeneous boundary conditions. Let this problem be given by

$$
\begin{aligned}
Lu &= g \quad \text{in } D, \\
\left(\frac{\partial}{\partial n} \right)^j u &= 0 \quad \text{on } \partial D, \; j = 0, \dots, m - 1.
\end{aligned}
\tag{1.2.5}
$$

We assume that the operator L is given in its *divergence form*

$$
Lu = \sum_{|\alpha| \le m} \sum_{|\beta| \le m} (-1)^{|\beta|} D^\beta (a_{\alpha, \beta} D^\alpha u).
\tag{1.2.6}
$$

For smooth coefficients a_α the divergence form always exists, see [54, Sec. 5.3]. We further assume $u \in C^{2m}(D) \cap C^m(\overline{D})$ to be a classical solution of (1.2.5). Let $v \in C_0^\infty(D)$ be a test function. Considering the L_2 scalar product, from (1.2.5) it follows

$$
\langle Lu, v \rangle_{L_2(D)} = \langle g, v \rangle_{L_2(D)} \quad \text{for all } v \in C_0^\infty(\Omega).
\tag{1.2.7}
$$

With integration by parts we obtain for the left-hand-side

$$\langle Lu, v \rangle_{L_2(D)} = \Big\langle \sum_{|\alpha|\leq m} \sum_{|\beta|\leq m} (-1)^{|\beta|} D^\beta(a_{\alpha,\beta} D^\alpha u), v \Big\rangle_{L_2(D)}$$

$$= \int_D \sum_{|\alpha|\leq m} \sum_{|\beta|\leq m} (-1)^{|\beta|} D^\beta(a_{\alpha,\beta}(x) D^\alpha u(x)) v(x) \, dx$$

$$= \sum_{|\alpha|\leq m} \sum_{|\beta|\leq m} (-1)^{|\beta|} \int_D D^\beta(a_{\alpha,\beta}(x) D^\alpha u(x)) v(x) \, dx$$

$$= \sum_{|\alpha|\leq m} \sum_{|\beta|\leq m} \int_D a_{\alpha,\beta}(x) D^\alpha u(x) D^\beta v(x) \, dx.$$

This expression defines a *bilinear form* $a(\cdot, \cdot)$ on $H_0^m(D) \times C_0^\infty(D)$:

$$a(u, v) := \sum_{|\alpha|\leq m} \sum_{|\beta|\leq m} \int_D a_{\alpha,\beta}(x) D^\alpha u(x) D^\beta v(x) \, dx. \tag{1.2.8}$$

The right-hand-side of (1.2.7) can be interpreted as a linear functional f:

$$f(v) := \langle g, v \rangle_{L_2(D)} = \int_D g(x) v(x) \, dx. \tag{1.2.9}$$

Since $C_0^\infty(D)$ is dense in $H_0^m(D)$, the bilinear form $a(\cdot, \cdot)$ can be extended to $H_0^m(D) \times H_0^m(D)$. This leads to the following definition.

Definition 1.9. The weak formulation of (1.2.5) is given by

$$a(u, v) = f(v) \quad \text{for all } v \in H_0^m(D). \tag{1.2.10}$$

A function $u \in H_0^m(D)$ that fulfils (1.2.10) is called *weak solution*. We note that for classical solutions (1.2.5) and (1.2.10) are equivalent since the integration by parts can be reversed. The bilinear form associated with the Poisson equation (1.2.4) is given by

$$a(u, v) = \int_D \langle \operatorname{grad} u(x), \operatorname{grad} v(x) \rangle \, dx. \tag{1.2.11}$$

Here, $\langle \cdot, \cdot \rangle$ denotes the standard scalar product in $\mathbb{R}^d$. To answer the question if (1.2.10) has a unique solution, we consider further properties of bilinear forms in general Hilbert spaces $\mathcal{H}$.

Definition 1.10. Let $\mathcal{H}$ be a Hilbert space. A bilinear form $a : \mathcal{H} \times \mathcal{H} \to \mathbb{R}$ is called *continuous*, if there exists a constant $C_s > 0$, such that it holds

$$a(u, v) \leq C_s \|u\|_{\mathcal{H}} \|v\|_{\mathcal{H}} \quad \text{for all } u, v \in \mathcal{H}. \tag{1.2.12}$$

A continuous bilinear form is called *elliptic*, is there exists a constant $C_e > 0$ such that it holds

$$a(u, u) \geq C_e \|u\|_{\mathcal{H}}^2 \quad \text{for all } u \in \mathcal{H}. \tag{1.2.13}$$

Lemma 1.11. *[54, Lem. 6.91] Let $a : \mathcal{H} \times \mathcal{H} \to \mathbb{R}$ be a continuous bilinear form with constant $C_s > 0$. There exists a bounded operator $A : \mathcal{H} \to \mathcal{H}'$ such that*

$$a(u, v) = (Au)(v) \quad \text{for all } u, v \in \mathcal{H}. \tag{1.2.14}$$

Furthermore, it holds that $|||A||| \leq C_s$.

Definition 1.12. Let $\mathcal{H} \subset \mathcal{G}$ be Hilbert spaces which are continuously and densely embedded. Identifying $\mathcal{G} = \mathcal{G}'$ leads to the inclusions

$$\mathcal{H} \subset \mathcal{G} \subset \mathcal{H}', \tag{1.2.15}$$

where $\mathcal{G} \subset \mathcal{H}'$ is continuously and densely embedded, too. (1.2.15) is called *Gelfand triple*. A continuous bilinear form $a : \mathcal{H} \times \mathcal{H} \to \mathbb{R}$ is called *coercive*, if there exist constants $C_e > 0$, $C_k \in \mathbb{R}$ such that it holds

$$a(u, u) \geq C_e \|u\|_{\mathcal{H}}^2 - C_k \|u\|_{\mathcal{G}}^2 \quad \text{for all } u \in \mathcal{H}. \tag{1.2.16}$$

We summarise conditions for the existence of unique solutions in the following theorem.

Theorem 1.13. *[54, Lem. 6.94, 6.97] Let one of the following conditions be fulfilled.*

(i) $a : \mathcal{H} \times \mathcal{H} \to \mathbb{R}$ is elliptic with constant C_e.

(ii) $a : \mathcal{H} \times \mathcal{H} \to \mathbb{R}$ is coercive with constants $C_e > 0$, $C_k \in \mathbb{R}$ and one of the following inequalities holds with $\varepsilon > 0$ or $\varepsilon' > 0$:

$$\inf_{u \in \mathcal{H}, \|u\|_{\mathcal{H}}=1} \left\{ \sup_{v \in \mathcal{H}, \|v\|_{\mathcal{H}}=1} |a(u, v)| \right\} = \varepsilon > 0, \tag{1.2.17}$$

$$\inf_{v \in \mathcal{H}, \|v\|_{\mathcal{H}}=1} \left\{ \sup_{u \in \mathcal{H}, \|u\|_{\mathcal{H}}=1} |a(u, v)| \right\} = \varepsilon' > 0. \tag{1.2.18}$$

Then, the associated operator A in (1.2.14) is boundedly invertible with $|||A^{-1}||| \leq \frac{1}{C_e}$ for elliptic $a(\cdot, \cdot)$ and $|||A^{-1}||| \leq \frac{1}{\varepsilon}$ for coercive $a(\cdot, \cdot)$.

Let us return to the case $\mathcal{H} = H_0^m(D)$. If the conditions of the preceding theorem are fulfilled, Lemma 1.11 implies that the weak formulation (1.2.10) has a unique solution

$$u = A^{-1}f. \tag{1.2.19}$$

It would be a mistake to assume that elliptic differential operators always induce elliptic bilinear forms. But, with some restrictions we are able to formulate conditions on L that lead to a uniquely solvable weak formulation. Firstly, from (1.2.8) one infers that for bounded coefficients $a_{\alpha,\beta} \in L_\infty(D)$ the bilinear form $a(\cdot, \cdot)$ is continuous. Secondly, the ellipticity is considered in the following theorems from [54, Sec. 7.2].

Theorem 1.14. *Let* $m = 1$, $a_{\alpha,\beta} \in L_\infty(D)$ *and* L *be uniformly elliptic. Then,* $a : H_0^1(D) \times H_0^1(D) \to \mathbb{R}$ *is coercive. Let further* $a_{\alpha,\beta} = 0$ *for* $|\alpha| + |\beta| \leq 1$. *Then,* $a(\cdot,\cdot)$ *is elliptic.*

Theorem 1.15. *Let* $a_{\alpha,\beta}$ *be constant for* $|\alpha| = |\beta| = m$ *and* L *be uniformly elliptic. Let further* $a_{\alpha,\beta} = 0$ *for* $0 < |\alpha| + |\beta| \leq 2m - 1$. *Then,* $a : H_0^m(D) \times H_0^m(D) \to \mathbb{R}$ *is elliptic.*

Theorem 1.16. *(Garding) Let* $a_{\alpha,\beta} \in L_\infty(D)$ *and* L *be uniformly elliptic. Let further* $a_{\alpha,\beta} \in C(\overline{D})$ *for* $|\alpha| = |\beta| = m$. *Then,* $a : H_0^m(D) \times H_0^m(D) \to \mathbb{R}$ *is coercive.*

Note that the Laplace operator fulfils the requirements to induce an elliptic bilinear form. Obviously it is uniformly elliptic, moreover the nonzero coefficients of the associated bilinear form (1.2.8) are given by $a_{\alpha,\beta} = 1$ for $\alpha = \beta$, $|\alpha| = 1$. Applying Theorem 1.14 immediately yields the ellipticity.

1.3 Frames

In this section we introduce the concept of frames. Frames are stable, redundant generating systems in Hilbert spaces. They represent an alternative approach of decomposing elements into certain building blocks. In contrast to Riesz bases, these decompositions are not unique. On the other hand, frames are easier to construct in certain cases of underlying domains. A detailed introduction to frames is given in [15]. Furthermore, we refer to [14,52].

Definition 1.17. Let $\mathcal{H}$ be a Hilbert space with inner product $\langle \cdot, \cdot \rangle$. Let I be a countable index set. The collection $\mathcal{F} = \{f_i\}_{i \in I} \subset \mathcal{H}$ is called *frame*, if there exist $A, B > 0$ such that:

$$A\|f\|_{\mathcal{H}}^2 \leq \sum_{i \in I} |\langle f, f_i \rangle|^2 \leq B\|f\|_{\mathcal{H}}^2 \quad \text{for all } f \in \mathcal{H}. \tag{1.3.1}$$

The optimal constants A, B are called *frame bounds*. The collection $\{f_i\}_{i \in I}$ is called *Bessel system*, if the upper inequality in (1.3.1) is fulfilled.

Our aim is to utilise a frame for certain investigations on elements of $\mathcal{H}$. To this end it is necessary to express elements $f \in \mathcal{H}$ in terms of frame elements, that means to gain a sequence $c \in \ell_2(I)$ representing the element f. On the other hand, for a given sequence c, one has to ensure that the resulting composition of frame elements converges to some element in $\mathcal{H}$. We formalise these operations. The *synthesis operator* $T_{\mathcal{F}}$ is defined by

$$\begin{aligned} T_{\mathcal{F}} &: \ell_2(I) \to \mathcal{H}, \\ \{c_i\}_{i \in I} &\mapsto \sum_{i \in I} c_i f_i. \end{aligned} \tag{1.3.2}$$

The *analysis operator* $T_{\mathcal{F}}^*$ is defined by

$$
\begin{aligned}
T_{\mathcal{F}}^* &: \mathcal{H} \to \ell_2(I), \\
f &\mapsto \{\langle f, f_i\rangle\}_{i\in I}.
\end{aligned}
\tag{1.3.3}
$$

The *frame operator* $S_{\mathcal{F}} = T_{\mathcal{F}} T_{\mathcal{F}}^*$ is defined by

$$
\begin{aligned}
S_{\mathcal{F}} &: \mathcal{H} \to \mathcal{H}, \\
f &\mapsto \sum_{i\in I} \langle f, f_i\rangle f_i.
\end{aligned}
\tag{1.3.4}
$$

If there is no danger of confusion about the frame, we abbreviate the notation and write simply T, T^*, S, respectively. The frame operator is invertible and self-adjoint, cf. [15, Lem. 5.1.5]. With this at hand and $f = S S^{-1} f$ we are able give the frame decomposition of an element $f \in \mathcal{H}$:

$$
f = \sum_{i\in I} \langle S^{-1}f, f_i\rangle f_i = \sum_{i\in I} \langle f, S^{-1}f_i\rangle f_i = \sum_{i\in I} \langle f, f_i\rangle S^{-1}f_i, \quad f \in \mathcal{H}.
\tag{1.3.5}
$$

The collection $\{S^{-1}f_i\}_{i\in I}$ is called *canonical dual frame*. More general, frames $\{g_i\}_{i\in I}$, $\{f_i\}_{i\in I}$ are called *dual frames*, if it holds

$$
f = \sum_{i\in I} \langle f, g_i\rangle f_i = \sum_{i\in I} \langle f, f_i\rangle g_i \quad \text{for all } f \in \mathcal{H}.
\tag{1.3.6}
$$

Definition 1.18. The collection $\{f_i\}_{i\in I} \subset \mathcal{H}$ is called *Riesz basis* if it holds

$$
\mathcal{H} = \overline{\operatorname{span}\{f_i, i \in I\}},
\tag{1.3.7}
$$

and there exist $A, B > 0$, such that for all finite sequences $c = \{c_i\}$ it holds that

$$
A\|c\|_{\ell_2}^2 \le \Big\| \sum_{i\in I} c_i f_i \Big\|^2 \le B\|c\|_{\ell_2}^2.
\tag{1.3.8}
$$

Here, the optimal A, B are called *Riesz bounds*. Note that every Riesz basis is a frame. Furthermore, we give two alternative characterisations for frames and Bessel sequences, respectively.

Proposition 1.19. *[15, Thm. 3.1.3.] The collection $\{f_i\}_{i\in I}$ is a Bessel sequence in $\mathcal{H}$, if and only if the synthesis operator T is bounded. Then it holds $B \ge |||T|||^2$.*

Proposition 1.20. *[91, Prop. 2.2] The collection $\{f_i\}_{i\in I}$ is a frame for $\mathcal{H}$, if $\mathcal{H} = \overline{\operatorname{span}\{f_i, i \in I\}}$ and there exist $A, B > 0$ such that*

$$
B^{-1}\|f\|_{\mathcal{H}}^2 \le \inf_{c\in\ell_2(I):Tc=f} \|c\|_{\ell_2(I)}^2 \le A^{-1}\|f\|_{\mathcal{H}}^2 \quad \text{for all } f \in \mathcal{H}.
\tag{1.3.9}
$$

The following propositions state some facts about the union of Bessel systems, frames and Riesz bases, cf. [26].

Proposition 1.21. *Let $\mathcal{H}$ be a Hilbert space. Then, it holds:*

(i) *The union of finitely many Bessel systems for $\mathcal{H}$ is a Bessel system for $\mathcal{H}$.*

(ii) *A frame for $\mathcal{H}$ united with a Bessel system for $\mathcal{H}$ is a frame for $\mathcal{H}$.*

(iii) *A Bessel system for $\mathcal{H}$ which includes a Riesz basis for $\mathcal{H}$ is a frame for $\mathcal{H}$.*

Proof. To prove (i), we assume $\mathcal{B}_j = \{b_i\}_{i \in I_j} \subset \mathcal{H}$, $j = 1, \ldots, n$, to be Bessel systems for $\mathcal{H}$ with Bessel bounds $B_j > 0$, $j = 1, \ldots, n$. Let $\mathcal{B} = \bigcup_{j=1}^{n} \mathcal{B}_j$ and $I = \bigcup_{j=1}^{n} I_j$. Then, we have

$$\| \{\langle f, b_i \rangle_{\mathcal{H}}\}_{i \in I} \|_{\ell_2(I)}^2 \leq \sum_{j=1}^{n} \| \{\langle f, b_i \rangle_{\mathcal{H}}\}_{i \in I_j} \|_{\ell_2(I_j)}^2 \leq \sum_{j=1}^{n} B_j \|f\|_{\mathcal{H}}^2.$$

For (ii) we assume $\mathcal{F} = \{f_i\}_{i \in I_1}$ and $\mathcal{B} = \{f_i\}_{i \in I_2}$ to be a frame and a Bessel system for $\mathcal{H}$, respectively. As every frame is a Bessel system, the right-hand inequality follows immediately from (i). For the left-hand inequality we write

$$\| \{\langle f, f_i \rangle_{\mathcal{H}}\}_{i \in I_1 \cup I_2} \|_{\ell_2(I_1 \cup I_2)}^2 \geq \| \{\langle f, f_i \rangle_{\mathcal{H}}\}_{i \in I_1} \|_{\ell_2(I_1)}^2 \geq A \|f\|_{\mathcal{H}}^2,$$

with $A > 0$ a lower frame bound of $\mathcal{F}$. For the proof of part (iii) we consider a Bessel system $\mathcal{B} = \{b_i\}_{i \in I}$ for $\mathcal{H}$ which contains a Riesz basis $\mathcal{R} = \{b_i\}_{i \in I_R}$ for $\mathcal{H}$. We only have to show the left-hand side inequality in (1.3.1). We write

$$\| \{\langle f, b_i \rangle_{\mathcal{H}}\}_{i \in I} \|_{\ell_2(I)}^2 \geq \| \{\langle f, b_i \rangle_{\mathcal{H}}\}_{i \in I_R} \|_{\ell_2(I_R)}^2 \geq A \|f\|_{\mathcal{H}}^2,$$

with $A > 0$ a lower Riesz bound of $\mathcal{R}$. To perform the last estimate we used the fact that every Riesz basis is also a frame. $\square$

To conclude this section we state a proposition which considers the image of frames, Bessel systems and Riesz bases under certain operators.

Proposition 1.22. *Let $\mathcal{H}_1$ and $\mathcal{H}_2$ be Hilbert spaces and $U : \mathcal{H}_1 \mapsto \mathcal{H}_2$ an operator. Then, it holds:*

(i) *If $\mathcal{B}$ is a Bessel system for $\mathcal{H}_1$ and U is bounded, then $U\mathcal{B}$ is a Bessel system for $\mathcal{H}_2$.*

(ii) *If $\mathcal{F}$ is a frame for $\mathcal{H}_1$ and U is bounded and surjective, then $U\mathcal{F}$ is a frame for $\mathcal{H}_2$.*

(iii) *If $\mathcal{R}$ is a Riesz bases for $\mathcal{H}_1$ and U is bounded and invertible, then $U\mathcal{R}$ is a Riesz basis for $\mathcal{H}_2$.*

Proof. At first, we assume that U is bounded and $\mathcal{B} = \{b_i\}_{i \in I}$ is a Bessel system for $\mathcal{H}_1$. For $g \in \mathcal{H}_2$, it is

$$\| \{\langle g, Ub_i \rangle_{\mathcal{H}_2}\}_{i \in I} \|_{\ell_2(I)}^2 = \| \{\langle U^*g, b_i \rangle_{\mathcal{H}_1}\}_{i \in I} \|_{\ell_2(I)}^2 \leq B \|U^*g\|_{\mathcal{H}_1}^2 \leq B \|U\|_{\mathcal{H}_1 \mapsto \mathcal{H}_2}^2 \|g\|_{\mathcal{H}_2}^2.$$

For the last inequality we used $\|U\|_{\mathcal{H}_1 \mapsto \mathcal{H}_2} = \|U^*\|_{\mathcal{H}_2 \mapsto \mathcal{H}_1}$. For a proof of part (ii) we refer to [15, Cor. 5.3.2]. To show (iii) we use the fact, that a system $\mathcal{R} = \{r_i\}_{i \in I}$ is a Riesz basis for a Hilbert space $\mathcal{H}$ if and only if there exists a Hilbert space $\mathcal{K}$ with an orthonormal basis $\{e_i\}_{i \in I}$ and a bounded and invertible operator $V : \mathcal{K} \mapsto \mathcal{H}$, such that $\mathcal{R} = \{Ve_i\}_{i \in I}$, cf. [15, Def. 3.6.1]. So let $\mathcal{R} = \{r_i\}_{i \in I}$ be a Riesz basis for $\mathcal{H}_1$ and U bounded and invertible. As mentioned above, $\mathcal{R}$ can be written as $\{Ve_i\}_{i \in I}$, with $V : \mathcal{K} \mapsto \mathcal{H}_1$ bounded and invertible. The composition $UV : \mathcal{K} \mapsto \mathcal{H}_2$ is bounded and invertible as well. Thus, the system $U\mathcal{R} = \{UVe_i\}_{i \in I}$ is a Riesz basis for $\mathcal{H}_2$. $\qquad \square$

Chapter 2

Wavelets

This chapter is dedicated to wavelets, in particular we discuss the construction of biorthogonal wavelet bases. In the first section we consider the classical case of wavelet bases on the real line, in the second section we consider wavelet type bases on the unit interval. For both cases we recall the abstract construction principles and introduce the special case of spline wavelet bases. Wavelets can be employed in several fields such as signal analysis or the numerical treatment of partial differential equations. The latter will be explained in the course of this thesis. There is an overwhelming amount of literature concerning wavelets, for an introduction we refer to [16,92]. A prominent textbook on wavelets is [41], for applications of wavelets let us mention [46,70].

2.1 Wavelets on the Real Line

We start with a brief introduction and restrict ourselves to the univariate case. *Wavelets* are certain functions that arise from dilation, translation and scaling of a single function. The collection of these functions spans the space $L_2(\mathbb{R})$ and is called *wavelet basis*. The first mention of a wavelet basis was made by A. Haar in [53]. The so-called Haar wavelet is the simplest example of a wavelet in $L_2(\mathbb{R})$, though it has nice properties such as compact support and orthogonality. It was implicitly constructed with the help of a refinable generator, namely the indicator function $\chi_{[0,1)}$. Figure 2.1 shows the mother wavelet $\psi := \chi_{[0,\frac{1}{2})} - \chi_{[\frac{1}{2},1)}$ and a dilated and translated copy. Further progress in wavelet theory has been made in the 1980s among others by Y. Meyer [61] and S. Mallat [60]. Finally, compactly supported wavelets with arbitrarily high smoothness were constructed by I. Daubechies in [40] for the first time. Our construction is based on biorthogonal wavelet bases, this approach was presented in [21]. We start with a systematic definition including stability and biorthogonality.

Definition 2.1. A Riesz basis Ψ of $L_2(\mathbb{R})$, given by

$$\Psi := \{\psi_{j,k} : j, k \in \mathbb{Z}\}, \tag{2.1.1}$$

$$\psi_{j,k} := 2^{j/2}\psi(2^j \cdot -k), \tag{2.1.2}$$

is called *wavelet basis*. The function $\psi : \mathbb{R} \to \mathbb{R}$ is called *mother wavelet*.
Let $\widetilde{\Psi} := \left\{ \widetilde{\psi}_{j,k}, j, k \in \mathbb{Z} \right\}$ denote another wavelet basis. Ψ und $\widetilde{\Psi}$ are called *biorthogonal*, if it holds

$$\langle \psi_{j,k}, \widetilde{\psi}_{j',k'} \rangle_{L_2(\mathbb{R})} = \delta_{j,j'} \delta_{k,k'}, \quad j, j', k, k' \in \mathbb{Z}. \tag{2.1.3}$$

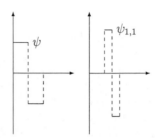

Figure 2.1: Haar wavelets.

In the sequel Ψ is called *primal* wavelet basis, where $\widetilde{\Psi}$ is called *dual* wavelet basis. The main utility for the construction of wavelet bases is the so-called multiresolution analysis. It goes back to S. Mallat, cf. [60].

Definition 2.2. A *multiresolution analysis (MRA)* is a sequence of closed subspaces $\{V_j\}_{j \in \mathbb{Z}}$ of $L_2(\mathbb{R})$ with the properties

$$...V_j \subset V_{j+1}... \tag{2.1.4}$$

$$\overline{\bigcup_{j \in \mathbb{Z}} V_j} = L_2(\mathbb{R}), \tag{2.1.5}$$

$$\bigcap_{j \in \mathbb{Z}} = \{0\}, \tag{2.1.6}$$

$$f \in V_j \Leftrightarrow f(\cdot - k) \in V_j, \quad k \in \mathbb{Z}, \tag{2.1.7}$$

$$f \in V_j \Leftrightarrow f(2\cdot) \in V_{j+1}. \tag{2.1.8}$$

Furthermore we require the space V_0 to be spanned by translates of a single function φ which is called *generator*:

$$V_0 = \overline{\mathrm{span}\{\varphi(\cdot - k), k \in \mathbb{Z}\}}. \tag{2.1.9}$$

The generator φ is called ℓ_2-*stable*, if there exist constants $A, B > 0$, such that it holds

$$A\|c\|_{\ell_2(\mathbb{Z})} \le \|\sum_{k \in \mathbb{Z}} c_k \varphi(\cdot - k)\|_{L_2(\mathbb{R})} \le B\|c\|_{\ell_2(\mathbb{Z})} \quad \text{for all } c \in \ell_2(\mathbb{Z}). \tag{2.1.10}$$

The concept of biorthogonality is carried over to multiresolution analyses. Let $\{V_j\}_{j\in\mathbb{Z}}$, $\{\widetilde{V}_j\}_{j\in\mathbb{Z}}$ be two MRAs. They are called biorthogonal, if it holds

$$\langle \varphi, \widetilde{\varphi}(\cdot - k)\rangle_{L_2(\mathbb{R})} = \delta_{0,k}, \quad k \in \mathbb{Z}. \tag{2.1.11}$$

The nestedness of the spaces V_j leads to the concept of refinable functions, which plays an important role in our considerations. A function f is called $(a,2)$-*refinable* if there exists a sequence $a \in \ell_1(\mathbb{Z})$ such that

$$f = \sum_{k\in\mathbb{Z}} a_k f(2\cdot -k). \tag{2.1.12}$$

Remark 2.3. It is a great convenience to consider the Fourier transform $\widehat{f}$ of a refinable function. It fulfils the equation

$$\widehat{f}(\xi) = \frac{1}{2}a(z)\widehat{f}\left(\frac{\xi}{2}\right), \quad \xi \in \mathbb{R}, z = e^{-i\frac{\xi}{2}} \in S_1, \tag{2.1.13}$$

where

$$a(z) := \sum_{k\in\mathbb{Z}} a_k z^k \tag{2.1.14}$$

is called the *symbol* of the function f.

Proof. We make use of the following property of the Fourier transform. For $k \in \mathbb{Z}$ it holds that

$$\widehat{f(2\cdot -k)}(\xi) = \frac{1}{2}\widehat{f}\left(\frac{\xi}{2}\right)e^{-i\frac{\xi}{2}k}. \tag{2.1.15}$$

Now let f fulfil (2.1.12). Due to the linearity of the Fourier transform and with (2.1.15) we obtain

$$\widehat{f}(\xi) = \sum_{k\in\mathbb{Z}} a_k \widehat{f(2\cdot -k)}(\xi)$$

$$= \frac{1}{2}\sum_{k\in\mathbb{Z}} a_k \widehat{f}\left(\frac{\xi}{2}\right)e^{-i\frac{\xi}{2}k}$$

$$= \frac{1}{2}\left(\sum_{k\in\mathbb{Z}} a_k e^{-i\frac{\xi}{2}k}\right)\widehat{f}\left(\frac{\xi}{2}\right).$$

With $z = e^{-i\frac{\xi}{2}}$, (2.1.13) follows. □

Let us return to the construction of wavelets based on two biorthogonal multiresolution analyses. We assume the generators to be refinable, hence it holds that

$$\widehat{\varphi}(\xi) = \frac{1}{2}a(z)\widehat{\varphi}\left(\frac{\xi}{2}\right), \quad \xi \in \mathbb{R}, z = e^{-i\frac{\xi}{2}} \in S_1, \tag{2.1.16}$$

$$\widehat{\widetilde{\varphi}}(\xi) = \frac{1}{2}\widetilde{a}(z)\widehat{\widetilde{\varphi}}\left(\frac{\xi}{2}\right), \quad \xi \in \mathbb{R}, z = e^{-i\frac{\xi}{2}} \in S_1. \tag{2.1.17}$$

The following theorems summarise the construction from [21].

Theorem 2.4. *Let $\varphi \in \mathscr{L}_2(\mathbb{R})$ be $(a, 2)$-refinable and let $\tilde{\varphi} \in \mathscr{L}_2(\mathbb{R})$ be $(\tilde{a}, 2)$-refinable. Let further (2.1.11) be fulfilled. Then it holds that*

$$a(z)\overline{\tilde{a}(z)} + a(-z)\overline{\tilde{a}(-z)} = 4, \quad z \in S_1. \tag{2.1.18}$$

Proof. We make use of the following property, cf. [16]. For functions $f, g \in L_2(\mathbb{R})$, $[f,g](z)$ is defined by

$$[f,g](z) := \sum_{k \in \mathbb{Z}} \langle f, g(\cdot - k) \rangle_{L_2(\mathbb{R})} z^k, \quad z \in S_1. \tag{2.1.19}$$

Then, for functions $f, g \in \mathscr{L}_2(\mathbb{R})$ it holds

$$[f,g](z) = \sum_{k \in \mathbb{Z}} \hat{f}(\xi + 2\pi k)\overline{\hat{g}(\xi + 2\pi k)}, \quad z = e^{-i\xi}. \tag{2.1.20}$$

Note that from (2.1.11) and (2.1.19) it immediately follows that

$$[\varphi, \tilde{\varphi}](z) = 1, \quad z \in S_1. \tag{2.1.21}$$

We start calculating using (2.1.20) and the refinability of $\varphi, \tilde{\varphi}$ as in (2.1.16), (2.1.17).

$$[\varphi, \tilde{\varphi}](e^{-i\xi}) = \sum_{k \in \mathbb{Z}} \hat{\varphi}(\xi + 2\pi k)\overline{\hat{\tilde{\varphi}}(\xi + 2\pi k)}$$

$$= \sum_{k \in \mathbb{Z}} \frac{1}{2} a(e^{-i\frac{\xi+2\pi k}{2}})\hat{\varphi}(\tfrac{\xi+2\pi k}{2})\overline{\frac{1}{2}\tilde{a}(e^{-i\frac{\xi+2\pi k}{2}})\hat{\tilde{\varphi}}(\tfrac{\xi+2\pi k}{2})}.$$

Splitting the sum yields

$$[\varphi, \tilde{\varphi}](e^{-i\xi}) = \sum_{k \in \mathbb{Z}} \frac{1}{4} a(e^{-i\frac{\xi+2\pi 2k}{2}})\hat{\varphi}(\tfrac{\xi+2\pi 2k}{2})\overline{\tilde{a}(e^{-i\frac{\xi+2\pi 2k}{2}})\hat{\tilde{\varphi}}(\tfrac{\xi+2\pi 2k}{2})}$$

$$+ \sum_{k \in \mathbb{Z}} \frac{1}{4} a(e^{-i\frac{\xi+2\pi(2k+1)}{2}})\hat{\varphi}(\tfrac{\xi+2\pi(2k+1)}{2})$$

$$\cdot \overline{\tilde{a}(e^{-i\frac{\xi+2\pi(2k+1)}{2}})\hat{\tilde{\varphi}}(\tfrac{\xi+2\pi(2k+1)}{2})}.$$

With the periodicity of the exponential function we get

$$[\varphi, \tilde{\varphi}](e^{-i\xi}) = \frac{1}{4} a(e^{-i\frac{\xi}{2}})\overline{\tilde{a}(e^{-i\frac{\xi}{2}})} \sum_{k \in \mathbb{Z}} \hat{\varphi}(\tfrac{\xi}{2} + 2\pi k)\overline{\hat{\tilde{\varphi}}(\tfrac{\xi}{2} + 2\pi k)}$$

$$+ \frac{1}{4} a(e^{-i(\frac{\xi}{2}+\pi)})\overline{\tilde{a}(e^{-i(\frac{\xi}{2}+\pi)})} \sum_{k \in \mathbb{Z}} \hat{\varphi}(\tfrac{\xi}{2} + \pi + 2\pi k)\overline{\hat{\tilde{\varphi}}(\tfrac{\xi}{2} + \pi + 2\pi k)}.$$

Again applying (2.1.20) leads to

$$[\varphi, \tilde{\varphi}](e^{-i\xi}) = \frac{1}{4} a(e^{-i\frac{\xi}{2}})\overline{\tilde{a}(e^{-i\frac{\xi}{2}})}[\varphi, \tilde{\varphi}](e^{-i\frac{\xi}{2}})$$

$$+ \frac{1}{4} a(e^{-i(\frac{\xi}{2}+\pi)})\overline{\tilde{a}(e^{-i(\frac{\xi}{2}+\pi)})}[\varphi, \tilde{\varphi}](e^{-i(\frac{\xi}{2}+\pi)}).$$

Setting $z = e^{-i\frac{\xi}{2}}$ and using (2.1.21) finally gives the claim. $\qquad\square$

We proceed with the construction process. One looks for bases of the spaces W_j, $\widetilde{W}_j$ which are defined by

$$V_{j+1} = V_j \oplus W_j, \qquad\qquad \widetilde{V}_{j+1} = \widetilde{V}_j \oplus \widetilde{W}_j, \qquad\qquad (2.1.22)$$

$$W_j \perp \widetilde{V}_j, \qquad\qquad \widetilde{W}_j \perp V_j. \qquad\qquad (2.1.23)$$

Then, it holds that

$$L_2(\mathbb{R}) = \bigoplus_{j \in \mathbb{Z}} W_j = V_0 \oplus \bigoplus_{j=0}^{\infty} W_j. \qquad\qquad (2.1.24)$$

From (2.1.4) and (2.1.8) it follows that is suffices to find bases for the spaces W_0, $\widetilde{W}_0$ which consist only of translates of single functions ψ, $\widetilde{\psi}$. Then, the spaces W_j, $\widetilde{W}_j$ are spanned by translates $\psi(2^j \cdot - k)$, $\widetilde{\psi}(2^j \cdot - k)$, respectively. Moreover, these functions have decompositions into fine generators:

$$\psi = \sum_{k \in \mathbb{Z}} b_k \varphi(2 \cdot - k), \quad b \in \ell_1(\mathbb{Z}), \qquad\qquad (2.1.25)$$

$$\widetilde{\psi} = \sum_{k \in \mathbb{Z}} \widetilde{b}_k \widetilde{\varphi}(2 \cdot - k), \quad \widetilde{b} \in \ell_1(\mathbb{Z}). \qquad\qquad (2.1.26)$$

Analogously to (2.1.16), (2.1.17) we have

$$\widehat{\psi}(\xi) = \frac{1}{2} b(z) \widehat{\varphi}\left(\frac{\xi}{2}\right), \quad z = e^{-i\frac{\xi}{2}}, \xi \in \mathbb{R}, \qquad\qquad (2.1.27)$$

$$\widehat{\widetilde{\psi}}(\xi) = \frac{1}{2} \widetilde{b}(z) \widehat{\widetilde{\varphi}}\left(\frac{\xi}{2}\right), \quad z = e^{-i\frac{\xi}{2}}, \xi \in \mathbb{R}. \qquad\qquad (2.1.28)$$

Here, $b(z)$ is called symbol of the wavelet ψ.

Theorem 2.5. *Let $\varphi \in \mathscr{L}_2(\mathbb{R})$ be $(a, 2)$-refinable and ℓ_2-stable and let $\widetilde{\varphi} \in \mathscr{L}_2(\mathbb{R})$ be $(\widetilde{a}, 2)$-refinable and ℓ_2-stable. Let further (2.1.11) be fulfilled. Let the wavelets ψ, $\widetilde{\psi}$ be defined by their symbols*

$$b(z) := -z \overline{\widetilde{a}(-z)}, \quad \widetilde{b}(z) := -z \overline{a(-z)}. \qquad\qquad (2.1.29)$$

Then, the systems $\{2^{j/2} \psi(2^j \cdot - k)\}_{j,k \in \mathbb{Z}}$, $\{2^{j/2} \widetilde{\psi}(2^j \cdot - k)\}_{j,k \in \mathbb{Z}}$ are biorthogonal wavelet bases of $L_2(\mathbb{R})$.

Proof. We begin with an auxiliary statement. From (2.1.29) and (2.1.18) it follows that the matrix $\begin{pmatrix} a(z) & a(-z) \\ b(z) & b(-z) \end{pmatrix}$ is invertible on S_1 with determinant $4z$. Furthermore it holds

$$\begin{pmatrix} a(z) & a(-z) \\ b(z) & b(-z) \end{pmatrix} \begin{pmatrix} \overline{\widetilde{a}(z)} & \overline{\widetilde{b}(z)} \\ \overline{\widetilde{a}(-z)} & \overline{\widetilde{b}(-z)} \end{pmatrix} = 4I. \qquad\qquad (2.1.30)$$

Next we show that (2.1.30) implies the biorthogonality conditions (2.1.23) and (2.1.3).

$W_0 \perp \tilde{V}_0$: Let $k \in \mathbb{Z}$ be arbitrary. It suffices to show $\langle \psi, \tilde{\varphi}(\cdot - k) \rangle_{L_2(\mathbb{R})} = 0$. By Plancherels theorem we have

$$\langle \psi, \tilde{\varphi}(\cdot - k) \rangle_{L_2(\mathbb{R})} = \frac{1}{2\pi} \langle \hat{\psi}, \hat{\tilde{\varphi}}(\cdot - k) \rangle_{L_2(\mathbb{R})}$$

$$= \frac{1}{2\pi} \int_{\mathbb{R}} \hat{\psi}(\xi) \overline{\hat{\tilde{\varphi}}(\xi)} e^{-ik\xi} \, d\xi.$$

Inserting the two-scale-relations (2.1.16), (2.1.27) and splitting the range of integration yields for $z = e^{-i\frac{\xi}{2}}$

$$\langle \psi, \tilde{\varphi}(\cdot - k) \rangle_{L_2(\mathbb{R})} = \frac{1}{2\pi} \int_{\mathbb{R}} \frac{1}{4} b(z) \hat{\varphi}(\tfrac{\xi}{2}) \overline{\tilde{a}(z)} \overline{\hat{\tilde{\varphi}}(\tfrac{\xi}{2})} e^{ik\xi} \, d\xi$$

$$= \frac{1}{8\pi} \sum_{l \in \mathbb{Z}} \int_{[0,4\pi)+4\pi l} b(z) \overline{\tilde{a}(z)} \hat{\varphi}(\tfrac{\xi}{2}) \overline{\hat{\tilde{\varphi}}(\tfrac{\xi}{2})} e^{ik\xi} \, d\xi$$

$$= \frac{1}{8\pi} \sum_{l \in \mathbb{Z}} \int_{[0,4\pi)} b(z) \overline{\tilde{a}(z)} \hat{\varphi}(\tfrac{\xi}{2} + 2\pi l) \overline{\hat{\tilde{\varphi}}(\tfrac{\xi}{2} + 2\pi l)} e^{ik(\xi + 4\pi l)} \, d\xi.$$

Using the periodicity and (2.1.20),(2.1.21) leads to

$$\langle \psi, \tilde{\varphi}(\cdot - k) \rangle_{L_2(\mathbb{R})} = \frac{1}{8\pi} \int_{[0,4\pi)} b(z) \overline{\tilde{a}(z)} e^{ik\xi} \, d\xi.$$

Again, splitting the range of integrations leads to

$$\langle \psi, \tilde{\varphi}(\cdot - k) \rangle_{L_2(\mathbb{R})} = \frac{1}{8\pi} \int_{[0,2\pi)} b(z) \overline{\tilde{a}(z)} e^{ik\xi} \, d\xi + \frac{1}{8\pi} \int_{[2\pi,4\pi)} b(z) \overline{\tilde{a}(z)} e^{ik\xi} \, d\xi$$

$$= \frac{1}{8\pi} \int_{[0,2\pi)} b(z) \overline{\tilde{a}(z)} e^{ik\xi} \, d\xi$$

$$+ \frac{1}{8\pi} \int_{[0,2\pi)} b(e^{-i(\frac{\xi}{2}+\pi)}) \overline{\tilde{a}(e^{-i(\frac{\xi}{2}+\pi)})} e^{ik(\xi+2\pi)} \, d\xi$$

$$= \frac{1}{8\pi} \int_{[0,2\pi)} e^{ik\xi} (b(z) \overline{\tilde{a}(z)} + b(-z) \overline{\tilde{a}(-z)}) \, d\xi.$$

Finally, with (2.1.30) the claim follows.

$\widetilde{W}_0 \perp V_0$: This case is completely analogous to the previous one. With (2.1.30) we have $b(z) \overline{\tilde{a}(z)} + b(-z) \overline{\tilde{a}(-z)} = 0$ and accordingly $\tilde{\psi}, \varphi(\cdot - k)$ are orthogonal.

(2.1.3): By transformation arguments it suffices to show that $\langle \psi, \tilde{\psi}(\cdot - k) \rangle = \delta_{0,k}$. With the same arguments as above we calculate

$$\langle \psi, \tilde{\psi}(\cdot - k) \rangle_{L_2(\mathbb{R})} = \frac{1}{8\pi} \int_{[0,2\pi)} e^{ik\xi} (b(z) \overline{\tilde{b}(z)} + b(-z) \overline{\tilde{b}(-z)}) \, d\xi.$$

With (2.1.30) it follows

$$\langle \psi, \widetilde{\psi}(\cdot - k) \rangle_{L_2(\mathbb{R})} = \frac{1}{8\pi} \int\limits_{[0,2\pi)} 4e^{ik\xi} \, \mathrm{d}\xi = \frac{1}{2\pi} 2\pi \delta_{0,k}.$$

It still remains to prove that the wavelets span a non-trivial complement space W_0 in V_1. To this end it suffices to show that a fine generator $\varphi(2 \cdot -\rho) \in V_1, \rho = 0, 1$ can be decomposed into wavelets and generators on a coarse level. Again we start with an auxiliary result. Defining *sub-symbols* $a_\rho(z^2) := \sum_{k \in \mathbb{Z}} a_{2k+\rho} z^{2k}$, $b_\rho(z^2)$, from (2.1.29) it follows that the matrix $\begin{pmatrix} a_0(z^2) & a_1(z^2) \\ b_0(z^2) & b_1(z^2) \end{pmatrix} = \frac{1}{2} \begin{pmatrix} a(z) & a(-z) \\ b(z) & b(-z) \end{pmatrix} \begin{pmatrix} 1 & \frac{1}{z} \\ 1 & -\frac{1}{z} \end{pmatrix}$ is invertible with inverse $\begin{pmatrix} c_0(z^2) & d_0(z^2) \\ c_1(z^2) & d_1(z^2) \end{pmatrix}$. Next we show that this relation is sufficient to obtain a non-trivial complement. Assume that such a complement space exists, then we have a reconstruction property for fine generators:

$$\varphi(2 \cdot -\rho) = \sum_{k \in \mathbb{Z}} c_{\rho+2k} \varphi(\cdot - k) + \sum_{n \in \mathbb{Z}} d_{\rho+2n} \psi(\cdot - n). \qquad (2.1.31)$$

Applying the Fourier transform, with (2.1.15) we get

$$\left(\frac{1}{2} e^{-i\frac{\xi}{2}\rho} I \right) \widehat{\varphi} \left(\frac{\xi}{2} \right) = c_\rho(z^2) \widehat{\varphi}(\xi) + d_\rho(z^2) \widehat{\psi}(\xi), \quad \xi \in \mathbb{R}, z \in S_1.$$

With the refinability of φ and (2.1.27) we get

$$\left(\frac{1}{2} e^{-i\frac{\xi}{2}\rho} I \right) \widehat{\varphi} \left(\frac{\xi}{2} \right) = \frac{1}{2} c_\rho(z^2) a(z) \widehat{\varphi} \left(\frac{\xi}{2} \right) + \frac{1}{2} d_\rho(z^2) b(z) \widehat{\varphi} \left(\frac{\xi}{2} \right).$$

Hence, a sufficient condition for (2.1.31) is given by

$$z^\rho I = c_\rho(z^2) a(z) + d_\rho(z^2) b(z), \quad z \in S_1. \qquad (2.1.32)$$

With the identity $\begin{pmatrix} a_0(z^2) & a_1(z^2) \\ b_0(z^2) & b_1(z^2) \end{pmatrix} = \frac{1}{2} \begin{pmatrix} a(z) & a(-z) \\ b(z) & b(-z) \end{pmatrix} \begin{pmatrix} 1 & \frac{z}{z} \\ 1 & -\frac{1}{z} \end{pmatrix}$ this is equivalent to

$$z^\rho I = c_\rho(z^2) \left(\sum_{\widehat{\rho}=0,1} z^{\widehat{\rho}} a_{\widehat{\rho}}(z^2) \right) + d_\rho(z^2) \left(\sum_{\widehat{\rho}=0,1} z^{\widehat{\rho}} b_{\widehat{\rho}}(z^2) \right)$$
$$= \sum_{\widehat{\rho}=0,1} z^{\widehat{\rho}} \left(c_\rho(z^2) a_{\widehat{\rho}}(z^2) + d_\rho(z^2) b_{\widehat{\rho}}(z^2) \right).$$

Hence, with the above description of the inverse matrix (2.1.31) follows. $\qquad \square$

So far, the construction of biorthogonal wavelet bases reduces to the question whether there exists a biorthogonal pair of generators which are refinable and ℓ_2-stable. This is indeed the case, and for the remainder of this section we stick to the case of *spline* generators and *spline wavelets*, respectively. The *cardinal B-spline* $N_m, m \in \mathbb{N}$, can recursively be defined by

$$N_1 := \chi_{[0,1)}, \tag{2.1.33}$$

$$N_m := (N_{m-1} * N_1) = \int_0^1 N_{m-1}(\cdot - t) \, dt, \quad m \geq 2. \tag{2.1.34}$$

Cardinal B-splines inherit nice properties, we collect some of them in the following lemma.

Lemma 2.6. *[16] For $m \in \mathbb{N}, k \in \mathbb{Z}, x \in \mathbb{R}$ it holds that*

$$N_m(x) = \frac{1}{(m-1)!} \sum_{k=0}^m (-1)^k \binom{n}{k} (x-k)_+^{m-1}, \tag{2.1.35}$$

$$N_m(x) = \frac{x}{m-1} N_{m-1}(x) + \frac{m-x}{m-1} N_{m-1}(x-1), \quad m \geq 2, \tag{2.1.36}$$

$$N_m'(x) = N_{m-1}(x) - N_{m-1}(x-1), \quad m \geq 2, \tag{2.1.37}$$

$$\operatorname{supp} N_m = [0, m], \tag{2.1.38}$$

$$\sum_{k \in \mathbb{Z}} N_m(x-k) = 1, \tag{2.1.39}$$

$$N_m \in H^{m-\frac{1}{2}}(\mathbb{R}). \tag{2.1.40}$$

Furthermore, N_m is $(a, 2)$-refinable with

$$N_m = \sum_{k=0}^m 2^{1-m} \binom{m}{k} N_m(2 \cdot - k). \tag{2.1.41}$$

That means, the symbol of the cardinal B-spline generator is given by the Laurent polynomial $a(z) = \sum_{k=0}^m 2^{1-m} \binom{m}{k} z^k = 2^{1-m}(1+z)^m$. The following theorem states the existence of a suitable dual generator associated with the cardinal B-spline and hence of a biorthogonal spline wavelet basis, cf. [21].

Theorem 2.7. *Let $\varphi := N_m$ denote the cardinal B-spline of order m with symbol $a(z)$. Let further $m, \widetilde{m} \in \mathbb{N}$ be such that $\widetilde{m} \geq m$, $m + \widetilde{m}$ is even. Then, there exists* an ℓ_2-stable, $(\tilde{a}, 2)$-refinable function $\tilde{\varphi}$ such that

$$\tilde{\varphi} = \sum_{k=-\tilde{m}-\lfloor\frac{m}{2}\rfloor+1}^{\tilde{m}+\lceil\frac{m}{2}\rceil-1} \tilde{a}_k \tilde{\varphi}(2 \cdot -k), \tag{2.1.42}$$

$$\tilde{a}_k = \sum_{n=0}^{\frac{m+\tilde{m}}{2}-1} \sum_{l=0}^{2n} 2^{-\tilde{m}-2n}(-1)^{n+l} \binom{\tilde{m}}{k+\lfloor\frac{m}{2}\rfloor-l+n}\binom{\frac{m+\tilde{m}}{2}-1+n}{n}\binom{2n}{l}, \tag{2.1.43}$$

where (2.1.18) is fulfilled and $\tilde{a}$, $\tilde{\varphi}$ are compactly supported. Furthermore, the wavelets ψ, $\tilde{\psi}$ defined by their symbols (2.1.29) possess m $(\tilde{m})$ vanishing moments.

Concrete examples of B-spline generators and wavelets are shown in Figures 2.2 - 2.3.

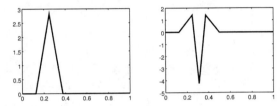

Figure 2.2: Primal CDF generator and wavelet for $m = \tilde{m} = 2$.

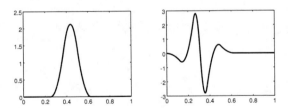

Figure 2.3: Primal CDF generator and wavelet for $m = \tilde{m} = 3$.

2.2 Wavelets on the Interval

When it comes to the numerical treatment of elliptic partial differential equations, one usually considers bounded domains. Hence the concept of wavelet bases has to be adapted. It turns out that it suffices to change the construction of those wavelets near the boundary. Strictly speaking, the resulting basis is of *wavelet type*. Since later on we use a tensor product ansatz, it suffices to construct wavelets on intervals, hence without loss of generality we can choose $I := [0, 1]$. At first we recall the abstract construction process. For details we refer to [35–39]. After that we discuss the particular basis developed by M. Primbs in [66, 67].

Definition 2.8. A dyadic multiresolution analysis of $L_2(I)$ is a sequence of closed subspaces $\{V_j\}_{j \in \mathbb{Z}, j \geq j_0}$ with the properties

$$\ldots V_{j_0} \subset V_{j_0+1} \ldots, \tag{2.2.1}$$

$$\bigcup_{j=j_0}^{\infty} V_j = L_2(I), \tag{2.2.2}$$

$$V_j = \mathrm{span}\{\varphi_{j,k} : k \in \Delta_j\}, \tag{2.2.3}$$

where $\Delta_j \subset \mathbb{Z}$ are finite index sets. Furthermore, the set of functions $\{\varphi_{j,k} : k \in \Delta_j\}$ is called ℓ_2-stable, if there exist constants $A, B > 0$, such that it holds

$$A\|c\|_{\ell_2(\Delta_j)} \leq \Big\| \sum_{k \in \Delta_j} c_{j,k}\varphi_{j,k} \Big\|_{L_2(I)} \leq B\|c\|_{\ell_2(\Delta_j)} \quad \text{for all } c \in \ell_2(\Delta_j). \tag{2.2.4}$$

Note that in contrast to the case of $L_2(\mathbb{R})$, we do not require the functions $\varphi_{j,k}$ to be translates of a single function. We drop the concept of $(a, 2)$-refinability, but from (2.2.1) we infer that the generators fulfil a *two-scale relation* of the form

$$\varphi_{j,k} = \sum_{l \in \Delta_{j+1}} a_{k,l}^j \varphi_{j+1,l}, \quad k \in \Delta_j, j \geq j_0. \tag{2.2.5}$$

Rearranging the functions in a vector $\Phi_j := \{\varphi_{j,k} : k \in \Delta_j\}$ and the two-scale coefficients in a matrix

$$M_{j,0} := (a_{k,l}^j)_{k \in \Delta_j, l \in \Delta_{j+1}}, \tag{2.2.6}$$

leads to the matrix equation

$$\Phi_j = M_{j,0}\Phi_{j+1}. \tag{2.2.7}$$

Let $\{\widetilde{V}_j\}_{j \geq j_0}$ denote a second multiresolution analysis of $L_2(I)$ with generators $\widetilde{\Phi}_j := \{\widetilde{\varphi}_{j,k} : k \in \Delta_j\}$ and two-scale matrix $\widetilde{M}_{j,0}$. Note that the index sets Δ_j coincide with the primal ones. $\{V_j\}_{j \geq j_0}$ and $\{\widetilde{V}_j\}_{j \geq j_0}$ are called biorthogonal, if it holds

$$\langle \varphi_{j,k}, \widetilde{\varphi}_{j,k'} \rangle_{L_2(I)} = \delta_{k,k'}, \quad k, k' \in \Delta_j, j \geq j_0. \tag{2.2.8}$$

One can rewrite the biorthogonality condition (2.2.8) as

$$I = \Phi_j \widetilde{\Phi}_j^T, \tag{2.2.9}$$

and hence

$$I = M_{j,0} \Phi_{j+1} \widetilde{\Phi}_{j+1}^T \widetilde{M}_{j,0}^T = M_{j,0} \widetilde{M}_{j,0}^T. \tag{2.2.10}$$

To clarify what we understand by a wavelet basis of $L_2(I)$, we consider the complement spaces W_j, $\widetilde{W}_j$ defined by (2.1.22), (2.1.23). With the definition

$$\nabla_j := \Delta_{j+1} \setminus \Delta_j, \tag{2.2.11}$$

we denote sets of functions

$$\Psi_j := \{\psi_{j,k} : k \in \nabla_j\}, \quad \widetilde{\Psi}_j := \{\widetilde{\psi}_{j,k} : k \in \nabla_j\}. \tag{2.2.12}$$

Definition 2.9. Let $\{V_j\}_{j \geq j_0}$, $\{\widetilde{V}_j\}_{j \geq j_0}$ be biorthogonal multiresolution analyses of $L_2(I)$. Let Ψ_j, $\widetilde{\Psi}_j$ be ℓ_2-stable bases of W_j, $\widetilde{W}_j$, respectively. If the sets

$$\Psi := \Phi_{j_0} \cup \bigcup_{j \geq j_0} \Psi_j, \quad \widetilde{\Psi} := \widetilde{\Phi}_{j_0} \cup \bigcup_{j \geq j_0} \widetilde{\Psi}_j, \tag{2.2.13}$$

are biorthogonal Riesz bases of $L_2(I)$, Ψ and $\widetilde{\Psi}$ are called biorthogonal *wavelet bases of* $L_2(I)$.

We briefly describe the construction of such a wavelet basis. Analogously to (2.1.25), we conclude that the primal wavelets fulfil a two-scale relation of the form

$$\psi_{j,k} = \sum_{l \in \Delta_{j+1}} b_{k,l}^j \varphi_{j+1,l}, \quad k \in \nabla_j, j \geq j_0. \tag{2.2.14}$$

An analogous result holds for the dual wavelets. Rearranging the coefficients in a matrix $M_{j,1} := (b_{k,l}^j)_{k \in \nabla_j, l \in \Delta_{j+1}}$ leads to the matrix equation

$$\Psi_j = M_{j,1} \Phi_{j+1}. \tag{2.2.15}$$

With two given multiresolution analyses, it remains so determine suitable matrices $M_{j,1}$ and $\widetilde{M}_{j,1}$. To derive a reconstruction sequence we get the further condition

$$\begin{pmatrix} \Phi_j \\ \Psi_j \end{pmatrix} = \begin{pmatrix} M_{j,0} \\ M_{j,1} \end{pmatrix} \Phi_{j+1}. \tag{2.2.16}$$

That means, the matrix $M_j := \begin{pmatrix} M_{j,0} \\ M_{j,1} \end{pmatrix}$ has to be invertible. From the biorthogonality conditions it follows

$$I = \begin{pmatrix} \Phi_j \\ \Psi_j \end{pmatrix} \begin{pmatrix} \widetilde{\Phi}_j \\ \widetilde{\Psi}_j \end{pmatrix}^T = M_j \Phi_{j+1} \widetilde{\Phi}_{j+1}^T \widetilde{M}_j^T = M_j \widetilde{M}_j^T.$$

To ensure Riesz stability one also requires $\|M_j\|_{\ell_2} \lesssim 1$, $\|\widetilde{M}_j\|_{\ell_2} \lesssim 1$. Accordingly, both the construction principle and the matrices $M_{j,1}$, $\widetilde{M}_{j,1}$ are called *stable completion*. Furthermore, [66, Prop. 5.1] states that the existence of the matrices $M_{j,1}$ and the stability of Ψ_j are equivalent. To investigate whether such stable completions exist, Jackson and Bernstein estimates play a key role. Let $H^s(I)$ be defined as in Section 1.1.

A *Jackson estimate* for the spaces V_j and $s \in \mathbb{R}$ is fulfilled if it holds that

$$\inf_{v \in V_j} \|f - v\|_{L_2(I)} \lesssim 2^{-js}\|f\|_{H^s(I)} \quad \text{for all } f \in H^s(I). \tag{2.2.17}$$

A *Bernstein estimate* for the spaces V_j and $s \in \mathbb{R}$ is fulfilled if it holds that

$$\|v\|_{H^s(I)} \lesssim 2^{js}\|v\|_{L_2(I)} \quad \text{for all } v \in V_j \cap H^s(I). \tag{2.2.18}$$

The following theorem summarises the conditions on two dual multiresolution analyses to ensure the existence of a biorthogonal wavelet basis on the interval. It summarises [66, Section 2.3], in particular Lemma 2.6, Lemma 2.14 and Lemma 2.16.

Theorem 2.10. *Let $\{V_j\}_{j \geq j_0}$, $\{\widetilde{V}_j\}_{j \geq j_0}$ be biorthogonal multiresolution analyses. Let further hold that*

$$\|\varphi_{j,k}\|_{L_2(I)} \lesssim 1, \quad k \in \Delta_j, j \geq j_0, \tag{2.2.19}$$
$$\#\{k \in \Delta_j : \operatorname{supp}\varphi_{j,k} \cap \operatorname{supp}\varphi_{j,k'} \neq \emptyset\} \lesssim 1 \quad \text{for all } k' \in \Delta_j. \tag{2.2.20}$$

Let additionally (2.2.19) and (2.2.20) be fulfilled for the dual MRA. Let further (2.2.17) and (2.2.18) with s_1 for the primal MRA and with s_2 for the dual MRA be fulfilled. Then, there exist biorthogonal Riesz wavelet bases Ψ, $\widetilde{\Psi}$ of $L_2(I)$ as defined in (2.2.13), (2.2.15). Furthermore, rescaled versions of Ψ, $\widetilde{\Psi}$ are Riesz bases for the Sobolev spaces $H^s(I)$, $s < \min\{s_1, s_2\}$.

Note that in general a stable completion $M_{j,1}$ is not unique. To obtain biorthogonality, an initial stable completion $\check{M}_{j,1}$ among all stable completions is chosen. Then,

$$M_{j,1} := (I - M_{j,0}^T \widetilde{M}_{j,0})\check{M}_{j,1} \tag{2.2.21}$$

is a stable completion, too. Similar to the case of $L_2(\mathbb{R})$, the construction of wavelets reduces to the construction of suitable MRAs.

Since our quarklet frames on the interval are based on the wavelet basis constructed by M. Primbs in [66, 67], we briefly describe this particular construction. The underlying generators also are B-splines, but generalise the ones from Section 2.1. The index sets $\Delta_j \subset \mathbb{Z}$ are chosen as

$$\Delta_j := \{-m + 1, \ldots, 2^j - 1\}. \tag{2.2.22}$$

Given the knots

$$t_k^j := \begin{cases} 0, & k = -m+1, \ldots, 0, \\ 2^{-j}k, & k = 1, \ldots, 2^j - 1, \\ 1, & k = 2^j, \ldots, 2^j + m - 1, \end{cases}$$

with boundary knots of multiplicity m and single inner knots, the *Schoenberg B-Splines* $B_{j,k}^m$ are defined by

$$B_{j,k}^m(x) := (t_{k+m}^j - t_k^j)\,(\,\cdot\, - x)_+^{m-1}[t_k^j, \ldots, t_{k+m}^j], \quad k \in \Delta_j, x \in I, \tag{2.2.23}$$

where $f[t_0, \ldots, t_n]$ denotes the n-th *divided difference* of the function f. The primal generators of the Primbs basis are defined by

$$\varphi_{j,k} := 2^{j/2} B_{j,k}^m, \quad k \in \Delta_j. \tag{2.2.24}$$

Similar to Lemma 2.6, we collect some properties of the Schoenberg B-Splines.

Lemma 2.11. *[66] For $m \in \mathbb{N}, j \in \mathbb{N}, k \in \Delta_j, x \in I$ it holds that*

$$B_{j,k}^m(x) = N_m(2^j x - k), \quad k = 0, \ldots, 2^j - m, \tag{2.2.25}$$

$$B_{j,k}^m(x) = B_{j,2^j-m-k}^m(1 - x), \quad k = -m+1, \ldots, 2^j - 1, \tag{2.2.26}$$

$$\operatorname{supp} B_{j,k}^m = [t_k^j, t_{k+m}^j], \tag{2.2.27}$$

$$\sum_{k \in \Delta_j} B_{j,k}^m(x) = 1. \tag{2.2.28}$$

Furthermore, the Schoenberg B-Splines fulfil a two-scale relation of the form (2.2.5), cf. [66, Lem. 3.15], and are included in the Sobolev spaces $H^s(I), s < m - \frac{1}{2}$. The following theorem summarises the construction from [66].

Theorem 2.12. *Let $m, \widetilde{m} \in \mathbb{N}$ be such that $\widetilde{m} \geq m$, $m + \widetilde{m}$ is even. The Schoenberg B-Splines generate a MRA of $L_2(I)$. There exists a dual MRA such that $\{V_j\}_{j \geq j_0}$, $\{\widetilde{V}_j\}_{j \geq j_0}$ are biorthogonal and fulfil the assumptions of Theorem 2.10. Furthermore, the wavelets $\psi_{j,k}$, $\widetilde{\psi}_{j,k}$ possess m $(\widetilde{m})$ vanishing moments.*

Remark 2.13. Note that it is possible to incorporate boundary conditions in the construction of the MRAs and the wavelet bases, respectively. Let $\vec{\sigma} = (\sigma_l, \sigma_r) \in \{0, \lfloor s + 1/2 \rfloor\}^2$ denote the order of boundary conditions. For example, $\vec{\sigma} = (0, 1)$ denotes free boundary conditions at $x = 0$ and homogeneous Dirichlet boundary conditions at $x = 1$. Dealing with free boundary conditions, we set $\Delta_j := \Delta_{j,(0,0)}$. Consequently, $H_{\vec{\sigma}}^s(I)$ will denote the closed subspace of $H^s(I)$ corresponding to these very boundary conditions. On the other hand, the corresponding dual wavelets always should be of free boundary type. This choice is motivated as follows: In numerical applications the vanishing moment property is essential, and with free boundary conditions the dual generators have polynomial exactness $\widetilde{m} - 1$, which allows to construct primal wavelets with $\widetilde{m}$ vanishing moments. The index sets with respect to boundary conditions can be generalised as follows:

$$\Delta_{j,\vec{\sigma}} := \{-m + 1 + \operatorname{sgn}\sigma_l, \ldots, 2^j - 1 - \operatorname{sgn}\sigma_r\}, \qquad (2.2.29)$$

$$\nabla_{\vec{\sigma}}^R := \{(j,k) : j \in \mathbb{N}_0, j \geq j_0 - 1, k \in \nabla_{j,\vec{\sigma}}\}. \qquad (2.2.30)$$

The construction described above can be adapted to the case of incorporated boundary conditions. In particular, Jackson and Bernstein estimates hold true. For details we refer to [66]. With the special case of Schoenberg B-splines as primal generators we are able to specify the results from Theorem 2.10, in particular the weighted system

$$\Psi_{\vec{\sigma}}^s := 2^{-j_0 s}\Phi_{j_0} \cup \bigcup_{j \geq j_0} 2^{-js}\Psi_j, \quad 0 \leq s < m - \frac{1}{2}, \qquad (2.2.31)$$

is a Riesz basis of $H_{\vec{\sigma}}^s(I)$ with dual basis $\widetilde{\Psi}_{\vec{\sigma}}^s$, which is defined analogously.

To conclude this section we consider concrete examples of the primal wavelets constructed in [66]. Figures 2.4-2.5 show inner and left boundary wavelets.

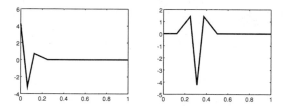

Figure 2.4: Primal Primbs wavelets for $m = \widetilde{m} = 2$.

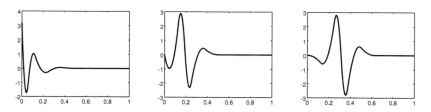

Figure 2.5: Primal Primbs wavelets for $m = \widetilde{m} = 3$.

For $m = \widetilde{m} = 2$, the matrices $M_{2,0}$, $M_{2,1}$ are given by

$$
M_{2,0} = \sqrt{2}
\begin{pmatrix}
1 & 0 & 0 & 0 & 0 \\
\frac{1}{2} & \frac{1}{2} & 0 & 0 & 0 \\
0 & 1 & 0 & 0 & 0 \\
0 & \frac{1}{2} & \frac{1}{2} & 0 & 0 \\
0 & 0 & 1 & 0 & 0 \\
0 & 0 & \frac{1}{2} & \frac{1}{2} & 0 \\
0 & 0 & 0 & 1 & 0 \\
0 & 0 & 0 & \frac{1}{2} & \frac{1}{2} \\
0 & 0 & 0 & 0 & 1
\end{pmatrix}, \quad
M_{2,1} = \sqrt{2}
\begin{pmatrix}
\frac{3}{2} & 0 & 0 & 0 \\
-\frac{9}{8} & \frac{1}{4} & 0 & 0 \\
\frac{1}{4} & \frac{1}{2} & 0 & 0 \\
\frac{1}{8} & -\frac{3}{2} & \frac{1}{4} & 0 \\
0 & \frac{1}{2} & \frac{1}{2} & 0 \\
0 & \frac{1}{4} & -\frac{3}{2} & \frac{1}{8} \\
0 & 0 & \frac{1}{2} & \frac{1}{4} \\
0 & 0 & \frac{1}{4} & -\frac{9}{8} \\
0 & 0 & 0 & \frac{3}{2}
\end{pmatrix}.
$$

For $m = \widetilde{m} = 3$, the matrices $M_{3,0}$, $M_{3,1}$ are given by

$$
M_{3,0} = \sqrt{2}
\begin{pmatrix}
1 & 0 & 0 & 0 & 0 & 0 & 0 & 0 & 0 & 0 \\
\frac{1}{2} & \frac{1}{2} & 0 & 0 & 0 & 0 & 0 & 0 & 0 & 0 \\
0 & \frac{3}{4} & \frac{1}{4} & 0 & 0 & 0 & 0 & 0 & 0 & 0 \\
0 & \frac{1}{4} & \frac{3}{4} & 0 & 0 & 0 & 0 & 0 & 0 & 0 \\
0 & 0 & \frac{3}{4} & \frac{1}{4} & 0 & 0 & 0 & 0 & 0 & 0 \\
0 & 0 & \frac{1}{4} & \frac{3}{4} & 0 & 0 & 0 & 0 & 0 & 0 \\
0 & 0 & 0 & \frac{3}{4} & \frac{1}{4} & 0 & 0 & 0 & 0 & 0 \\
0 & 0 & 0 & \frac{1}{4} & \frac{3}{4} & 0 & 0 & 0 & 0 & 0 \\
0 & 0 & 0 & 0 & \frac{3}{4} & \frac{1}{4} & 0 & 0 & 0 & 0 \\
0 & 0 & 0 & 0 & \frac{1}{4} & \frac{3}{4} & 0 & 0 & 0 & 0 \\
0 & 0 & 0 & 0 & 0 & \frac{3}{4} & \frac{1}{4} & 0 & 0 & 0 \\
0 & 0 & 0 & 0 & 0 & \frac{1}{4} & \frac{3}{4} & 0 & 0 & 0 \\
0 & 0 & 0 & 0 & 0 & 0 & \frac{3}{4} & \frac{1}{4} & 0 & 0 \\
0 & 0 & 0 & 0 & 0 & 0 & \frac{1}{4} & \frac{3}{4} & 0 & 0 \\
0 & 0 & 0 & 0 & 0 & 0 & 0 & \frac{3}{4} & \frac{1}{4} & 0 \\
0 & 0 & 0 & 0 & 0 & 0 & 0 & \frac{1}{4} & \frac{3}{4} & 0 \\
0 & 0 & 0 & 0 & 0 & 0 & 0 & 0 & \frac{1}{2} & \frac{1}{2} \\
0 & 0 & 0 & 0 & 0 & 0 & 0 & 0 & 0 & 1
\end{pmatrix},
$$

$$
M_{3,1} = \sqrt{2} \begin{pmatrix}
\frac{9}{8} & 0 & 0 & 0 & 0 & 0 & 0 & 0 \\
-\frac{327}{256} & -\frac{15}{32} & 0 & 0 & 0 & 0 & 0 & 0 \\
\frac{577}{1024} & -\frac{15}{128} & -\frac{3}{32} & 0 & 0 & 0 & 0 & 0 \\
\frac{51}{1024} & \frac{195}{128} & -\frac{9}{32} & 0 & 0 & 0 & 0 & 0 \\
-\frac{75}{512} & -\frac{75}{64} & \frac{7}{32} & -\frac{3}{32} & 0 & 0 & 0 & 0 \\
-\frac{13}{512} & -\frac{13}{64} & \frac{45}{32} & -\frac{9}{32} & 0 & 0 & 0 & 0 \\
\frac{27}{1024} & \frac{27}{128} & -\frac{45}{32} & \frac{7}{32} & \frac{3}{32} & 0 & 0 & 0 \\
\frac{9}{1024} & \frac{9}{128} & -\frac{7}{32} & \frac{45}{32} & \frac{9}{32} & 0 & 0 & 0 \\
0 & 0 & \frac{9}{32} & -\frac{45}{32} & -\frac{7}{32} & \frac{3}{32} & 0 & 0 \\
0 & 0 & \frac{3}{32} & -\frac{7}{32} & -\frac{45}{32} & \frac{9}{32} & 0 & 0 \\
0 & 0 & 0 & \frac{9}{32} & \frac{45}{32} & -\frac{7}{32} & \frac{9}{128} & \frac{9}{1024} \\
0 & 0 & 0 & \frac{3}{32} & \frac{7}{32} & -\frac{45}{32} & \frac{27}{128} & \frac{27}{1024} \\
0 & 0 & 0 & 0 & -\frac{9}{32} & \frac{45}{32} & -\frac{13}{64} & -\frac{13}{512} \\
0 & 0 & 0 & 0 & -\frac{3}{32} & \frac{7}{32} & -\frac{75}{64} & -\frac{75}{512} \\
0 & 0 & 0 & 0 & 0 & -\frac{9}{32} & \frac{195}{128} & \frac{51}{1024} \\
0 & 0 & 0 & 0 & 0 & -\frac{3}{32} & -\frac{15}{128} & \frac{577}{1024} \\
0 & 0 & 0 & 0 & 0 & 0 & -\frac{15}{32} & -\frac{327}{256} \\
0 & 0 & 0 & 0 & 0 & 0 & 0 & \frac{9}{8}
\end{pmatrix}.
$$

Chapter 3

Reconstruction of Multigenerators

In this chapter we establish a criterion for the reconstruction of general multigenerators. As we will see later, *reconstruction properties* of our quarklets are a nice utility. In the setting of Chapter 2 they allow to transform single-scale functions such as fine generators into multiscale functions such as wavelets and can be interpreted as the reversing of the two-scale relations (2.1.12), (2.1.25). Since the quarks are not refinable, we need to consider a more general setting. It turns out that vectors of quarklets fit in the setting of multiwavelets. Hence, in this chapter we develop an approach to obtain reconstruction properties of general multigenerators.

In the sequel we use Fourier techniques and restrict our discussion to the shift-invariant case. For an introduction to multiwavelets we refer to [33, 59, 65]. The findings of this chapter have been published in [31]. The main result is given in Theorem 3.2.

Let $\Phi = (\Phi_0, \ldots, \Phi_p)^T$, $p \in \mathbb{N}_0$ be a vector of functions from $L_1(\mathbb{R}) \cap L_2(\mathbb{R})$. The function vector Φ is called *refinable* if there exists a sequence of $(p+1) \times (p+1)$-matrices A_k such that

$$\Phi(x) = \sum_{k \in \mathbb{Z}} A_k \Phi(2x - k), \quad x \in \mathbb{R}, \ \{A_k\}_{k \in \mathbb{Z}} \in \ell_2(\mathbb{Z})^{(p+1) \times (p+1)}. \tag{3.1}$$

To avoid technical difficulties, in the sequel we will always assume the stronger condition $\{A_k\}_{k \in \mathbb{Z}} \in \ell_1(\mathbb{Z})^{(p+1) \times (p+1)}$. This more general concept of refinability allows us to employ techniques from wavelet analysis. Usually wavelets are constructed with the help of multiresolution analyses $\{V_j\}_{j \in \mathbb{Z}}$. Similarly, we define subspaces $V_{p,j}, j \in \mathbb{Z}, p \in \mathbb{N}_0$ of $L_2(\mathbb{R})$:

$$V_{p,j} := \overline{\mathrm{span}\{\Phi_q(2^j \cdot -k) : 0 \le q \le p, k \in \mathbb{Z}\}}^{L_2}, \tag{3.2}$$

It immediately follows that the sequence $V_{p,j}$ is nested both with respect to the level j and the parameter p. Defining a second function vector $\Psi = (\Psi_0, \ldots, \Psi_p)^T$ by

$$\Psi(x) := \sum_{k \in \mathbb{Z}} B_k \Phi(2x - k), \quad x \in \mathbb{R}, \tag{3.3}$$

where B_k are $(p+1) \times (p+1)$-matrices, the question arises whether the functions defined in (3.3) span an algebraic complement $W_{p,j}$ such that

$$V_{p,j+1} = V_{p,j} \oplus W_{p,j}. \tag{3.4}$$

We start with an auxiliary result, cf. [64].

Proposition 3.1.

(i) *The Fourier transform* $\widehat{\Phi} = (\widehat{\Phi}_0, \ldots, \widehat{\Phi}_p)^T$ *of a refinable function vector fulfils the matrix equation*

$$\widehat{\Phi}(\xi) = \frac{1}{2}\mathscr{A}(z)\widehat{\Phi}\left(\frac{\xi}{2}\right), \quad \xi \in \mathbb{R}, z = e^{-i\frac{\xi}{2}} \in S_1, \tag{3.5}$$

where

$$(\mathscr{A}(z))_{q,l} := \sum_{k \in \mathbb{Z}} (A_k)_{q,l} z^k \tag{3.6}$$

is called the symbol matrix *of* Φ.

(ii) *For a symbol matrix* $\mathscr{A}(z)$ *and* $\rho \in \{0,1\}$ *we define the* sub-symbol matrices $\mathscr{A}_\rho(z^2)$ *by*

$$\left(\mathscr{A}_\rho(z^2)\right)_{q,l} := \sum_{k \in \mathbb{Z}} (A_{2k+\rho})_{q,l} z^{2k}. \tag{3.7}$$

Then, it holds that

$$\mathscr{A}_0(z^2) = \frac{1}{2}\left(\mathscr{A}(z) + \mathscr{A}(-z)\right), \quad \mathscr{A}_1(z^2) = \frac{1}{2z}\left(\mathscr{A}(z) - \mathscr{A}(-z)\right). \tag{3.8}$$

Proof.

(i) Applying the Fourier transform to the q-th component of Φ, we obtain

$$\widehat{\Phi}_q(\xi) = \sum_{k \in \mathbb{Z}} \sum_{l=0}^{q} (A_k)_{q,l}\, \widehat{\Phi_l(2 \cdot - k)}(\xi)$$

$$= \sum_{k \in \mathbb{Z}} \sum_{l=0}^{q} (A_k)_{q,l}\, \frac{1}{2}\widehat{\Phi}_l\left(\frac{\xi}{2}\right) e^{-i\frac{\xi}{2}k}$$

$$= \frac{1}{2}\sum_{l=0}^{q} \underbrace{\left(\sum_{k \in \mathbb{Z}} (A_k)_{q,l}\, z^k\right)}_{:=(\mathscr{A}(z))_{q,l}} \widehat{\Phi}_l\left(\frac{\xi}{2}\right)$$

(ii) We consider an arbitrary entry of the matrices. Let $0 \leq q, l \leq p$. We calculate

$$
\begin{aligned}
\left(\frac{1}{2} \left(\mathscr{A}(z) + \mathscr{A}(-z) \right) \right)_{q,l} &= \frac{1}{2} \left(\sum_{k \in \mathbb{Z}} (A_k)_{q,l} z^k + \sum_{k \in \mathbb{Z}} (A_k)_{q,l} (-z)^k \right) \\
&= \frac{1}{2} \sum_{k \in \mathbb{Z}} 2(A_{2k})_{q,l} z^{2k} = \left(\mathscr{A}_0(z^2) \right)_{q,l},
\end{aligned}
$$

$$
\begin{aligned}
\left(\frac{1}{2z} \left(\mathscr{A}(z) - \mathscr{A}(-z) \right) \right)_{q,l} &= \frac{1}{2z} \left(\sum_{k \in \mathbb{Z}} (A_k)_{q,l} z^k - \sum_{k \in \mathbb{Z}} (A_k)_{q,l} (-z)^k \right) \\
&= \frac{1}{2z} \sum_{k \in \mathbb{Z}} 2(A_{2k+1})_{q,l} z^{2k+1} = \left(\mathscr{A}_1(z^2) \right)_{q,l}. \qquad \square
\end{aligned}
$$

Applying the Fourier transform to (3.3) in an analogous way, it follows that

$$
\widehat{\Psi}(\xi) = \frac{1}{2} \mathscr{B}(z) \widehat{\Phi} \left(\frac{\xi}{2} \right), \quad (\mathscr{B}(z))_{q,l} := \sum_{k \in \mathbb{Z}} (B_k)_{q,l} \, z^k. \tag{3.9}
$$

To achieve a decomposition relation, we need another preparatory result. By refinability, it suffices to consider $j = 1$. Because the spaces $V_{p,j}$ are shift-invariant and it holds $\Phi_p(2x - k) = \Phi_p(2(x - \tilde{k}) - \rho), \tilde{k} \in \mathbb{Z}, \rho \in \{0, 1\}$, it is sufficient to derive a decomposition relation of $\Phi_p(2x - \rho)$. The following theorem is the main result of this chapter.

Theorem 3.2. *Suppose that there exist $(p+1) \times (p+1)$ matrices $C_{\rho+2k}, D_{\rho+2k}$, such that*

$$
\begin{pmatrix} \mathscr{C}_0(z^2) & \mathscr{D}_0(z^2) \\ \mathscr{C}_1(z^2) & \mathscr{D}_1(z^2) \end{pmatrix} \begin{pmatrix} \mathscr{A}_0(z^2) & \mathscr{A}_1(z^2) \\ \mathscr{B}_0(z^2) & \mathscr{B}_1(z^2) \end{pmatrix} = I, \tag{3.10}
$$

where the sub-symbol matrices $\mathscr{C}_\rho(z^2)$, $\rho \in \{0, 1\}$, are defined by

$$
\left(\mathscr{C}_\rho(z^2) \right)_{q,l} = \sum_{k \in \mathbb{Z}} (C_{\rho+2k})_{q,l} z^{2k}, \quad \left(\mathscr{D}_\rho(z^2) \right)_{q,l} = \sum_{k \in \mathbb{Z}} (D_{\rho+2k})_{q,l} z^{2k}.
$$

Then, each function vector $\Phi(2 \cdot - \rho)$ has a decomposition in terms of wavelets Ψ and generators Φ on a coarse level, i.e., it holds

$$
\Phi(2x - \rho) = \sum_{k \in \mathbb{Z}} C_{\rho+2k} \Phi(x - k) + \sum_{n \in \mathbb{Z}} D_{\rho+2n} \Psi(x - n), \quad x \in \mathbb{R}. \tag{3.11}
$$

Proof. Applying the Fourier transform component-wise to (3.11) yields

$$
\left(\frac{1}{2} e^{-i \frac{\xi}{2} \rho} I \right) \widehat{\Phi} \left(\frac{\xi}{2} \right) = \mathscr{C}_\rho(z^2) \widehat{\Phi}(\xi) + \mathscr{D}_\rho(z^2) \widehat{\Psi}(\xi), \quad \xi \in \mathbb{R}, z \in S_1.
$$

With (3.5) and (3.9) we get

$$\left(\frac{1}{2}e^{-i\frac{\xi}{2}\rho}I\right)\widehat{\Phi}\left(\frac{\xi}{2}\right) = \frac{1}{2}\mathscr{C}_\rho(z^2)\mathscr{A}(z)\widehat{\Phi}\left(\frac{\xi}{2}\right) + \frac{1}{2}\mathscr{D}_\rho(z^2)\mathscr{B}(z)\widehat{\Phi}\left(\frac{\xi}{2}\right).$$

Hence, a sufficient condition for (3.11) is given by

$$z^\rho I = \mathscr{C}_\rho(z^2)\mathscr{A}(z) + \mathscr{D}_\rho(z^2)\mathscr{B}(z), \quad z \in S_1. \tag{3.12}$$

With (3.8) this is equivalent to

$$z^\rho I = \mathscr{C}_\rho(z^2)\left(\sum_{\widehat{\rho}=0,1} z^{\widehat{\rho}}\mathscr{A}_{\widehat{\rho}}(z^2)\right) + \mathscr{D}_\rho(z^2)\left(\sum_{\widehat{\rho}=0,1} z^{\widehat{\rho}}\mathscr{B}_{\widehat{\rho}}(z^2)\right)$$
$$= \sum_{\widehat{\rho}=0,1} z^{\widehat{\rho}}\left(\mathscr{C}_\rho(z^2)\mathscr{A}_{\widehat{\rho}}(z^2) + \mathscr{D}_\rho(z^2)\mathscr{B}_{\widehat{\rho}}(z^2)\right).$$

Hence, by (3.10) the claim follows. $\qquad\square$

The following proposition states that, with a transformation matrix $\mathscr{E}(z)$, both symbol matrices and sub-symbol matrices are convenient for the computation of refinement coefficients.

Proposition 3.3. *Defining*

$$X(z) := \frac{1}{2}\begin{pmatrix} \mathscr{A}(z) & \mathscr{A}(-z) \\ \mathscr{B}(z) & \mathscr{B}(-z) \end{pmatrix}, \quad \mathscr{E}(z) := \begin{pmatrix} I & \frac{1}{z}I \\ I & -\frac{1}{z}I \end{pmatrix}, \tag{3.13}$$

it holds that

$$\begin{pmatrix} \mathscr{A}_0(z^2) & \mathscr{A}_1(z^2) \\ \mathscr{B}_0(z^2) & \mathscr{B}_1(z^2) \end{pmatrix} = X(z)\mathscr{E}(z). \tag{3.14}$$

Moreover, $\mathscr{E}(z)$ is invertible on S_1 with

$$\det \mathscr{E}(z) = 2^{p+1}(-z)^{-p-1}, \quad \mathscr{E}(z)^{-1} = \frac{1}{2}\begin{pmatrix} I & I \\ zI & -zI \end{pmatrix}. \tag{3.15}$$

Proof. One easily verifies (3.14) with the help of (3.8):

$$X(z)\mathscr{E}(z) = \begin{pmatrix} \frac{1}{2}(\mathscr{A}(z) + \mathscr{A}(-z)) & \frac{1}{2z}(\mathscr{A}(z) - \mathscr{A}(-z)) \\ \frac{1}{2}(\mathscr{B}(z) + \mathscr{B}(-z)) & \frac{1}{2z}(\mathscr{B}(z) - \mathscr{B}(-z)) \end{pmatrix}$$
$$= \begin{pmatrix} \mathscr{A}_0(z^2) & \mathscr{A}_1(z^2) \\ \mathscr{B}_0(z^2) & \mathscr{B}_1(z^2) \end{pmatrix}.$$

Using block determinant formulas, cf. [78], we have

$$\det \mathscr{E}(z) = \det(-\frac{1}{z}I)\det(I + zI\frac{1}{z}II)$$
$$= \det(-\frac{1}{z}I)\det(2I)$$
$$= (-z)^{-p-1}2^{p+1}.$$

Additionally, one easily verifies $\mathscr{E}(z)\mathscr{E}(z)^{-1} = I$. $\qquad\square$

Remark 3.4. In case that the matrix

$$X(z)\mathscr{E}(z) = \begin{pmatrix} \mathscr{A}_0(z^2) & \mathscr{A}_1(z^2) \\ \mathscr{B}_0(z^2) & \mathscr{B}_1(z^2) \end{pmatrix}$$

in (3.10) is invertible on the torus S_1, it follows that all entries in

$$\begin{pmatrix} \mathscr{C}_0(z^2) & \mathscr{D}_0(z^2) \\ \mathscr{C}_1(z^2) & \mathscr{D}_1(z^2) \end{pmatrix}$$

consist of symbols whose coefficients are contained in $\ell_1(\mathbb{Z})$. Indeed, by our assumption, every entry in $X(z)\mathscr{E}(z)$ has this property, therefore the same holds for the determinant. Consequently, if the determinant does not vanish on the torus, the result follows by an application of the Wiener lemma, see [71, P. 278] for details.

Remark 3.5. In practice, one usually works with compactly supported generators that possess finitely supported masks $\{A_k\}_{k\in\mathbb{Z}}$. Then, it is of course desirable to find wavelets such that the entries of the matrix

$$\begin{pmatrix} \mathscr{A}(z) & \mathscr{A}(-z) \\ \mathscr{B}(z) & \mathscr{B}(-z) \end{pmatrix}^{-1}$$

consist of Laurent polynomials, for then the reconstruction sequences in (3.11) are also finitely supported. Fortunately, in the quarklet case, this is indeed the case, see Section 4.2.

Chapter 4

Univariate Quarklets

In this chapter we discuss quarklets in the univariate setting. We proceed in the following way. First we fix the underlying wavelet basis. In the shift invariant case we employ the CDF spline wavelet basis, cf. Section 2.1, in the interval case we employ the Primbs basis, cf. Section 2.2. After that we are able to define quarks similar to generators in the wavelet case. In Section 4.1 we study their basic properties. In Section 4.2 we consider quarklets in the shift-invariant case. With the results from Chapter 3 we derive reconstruction properties of the quarklets. Furthermore, it turns out that the boundary adapted quarklets need special treatment. This is done in Section 4.3. In particular the mask of the boundary quarklets has to be chosen appropriately. In Sections 4.5, 4.6 we show the stability of quarklet systems in scales of Besov spaces. From this result we easily infer the crucial frame property in Sobolev spaces. Since the cases of the real line and the interval are similar, we present a uniform treatment and point out the differences. A complete treatment of the shift-invariant case can be found in [28, 57].

4.1 Quarks: Definition and Basic Properties

In this section we introduce quarks and study some of their basic properties, in particular refinability, norm and Bernstein estimates. By a quark generator we understand the product of a suitable wavelet generator and a scaled monomial of degree p. We restrict our discussion to spline quarks.

Definition 4.1. Let $\varphi = N_m(\cdot + \lfloor \frac{m}{2} \rfloor)$ denote the *symmetrised* cardinal B-spline of order $m \in \mathbb{N}$, see (2.1.34). Then, the p-th *cardinal B-spline quark* φ_p is defined by

$$\varphi_p := \left(\frac{\cdot}{\lceil \frac{m}{2} \rceil} \right)^p N_m \left(\cdot + \lfloor \frac{m}{2} \rfloor \right), \quad p \in \mathbb{N}_0. \tag{4.1.1}$$

One could transfer this definition to other generators on the real line. Since we want to derive frames on bounded domains, we introduce quarks on the interval. As the underlying wavelet basis we choose the one constructed by M. Primbs.

Definition 4.2. Let Δ_j be as in (2.2.22) and $\varphi_{j,k}$ as in (2.2.24). Then, the p-th *Schoenberg B-spline quarks* $\varphi_{p,j,k}$ are defined by

$$
\varphi_{p,j,k} := \begin{cases}
\left(\dfrac{2^j \cdot}{k+m}\right)^p \varphi_{j,k}, & k = -m+1, ..., -1, \\[2ex]
\left(\dfrac{2^j \cdot -k - \lfloor \frac{m}{2} \rfloor}{\lceil \frac{m}{2} \rceil}\right)^p \varphi_{j,k}, & k = 0, ..., 2^j - m, \\[2ex]
\varphi_{p,j,2^j-m-k}(1 - \cdot), & k = 2^j - m + 1, ..., 2^j - 1.
\end{cases}
\tag{4.1.2}
$$

Figures 4.1-4.5 show left boundary and inner Schoenberg B-spline quarks. Note that the inner Schoenberg B-spline quarks are translated copies of the cardinal B-spline quark. With (2.2.25) we calculate for $k = 0, \ldots, 2^j - m$, $x \in [0,1]$:

$$
\begin{aligned}
\varphi_{p,j,k}(x) &= \left(\frac{2^j x - k - \lfloor \frac{m}{2} \rfloor}{\lceil \frac{m}{2} \rceil}\right)^p 2^{j/2} B_{j,k}^m(x) \\[2ex]
&= \left(\frac{2^j x - k - \lfloor \frac{m}{2} \rfloor}{\lceil \frac{m}{2} \rceil}\right)^p 2^{j/2} N_m(2^j x - k) \\[2ex]
&= \left(\frac{2^j x - k - \lfloor \frac{m}{2} \rfloor}{\lceil \frac{m}{2} \rceil}\right)^p 2^{j/2} N_m\left(2^j x - k - \lfloor \frac{m}{2} \rfloor + \lfloor \frac{m}{2} \rfloor\right) \\[2ex]
&= 2^{j/2} \varphi_p\left(2^j x - k - \lfloor \frac{m}{2} \rfloor\right).
\end{aligned}
$$

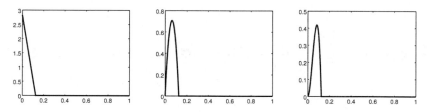

Figure 4.1: Left boundary quark for $m = 2$ and $p = 0, 1, 2$.

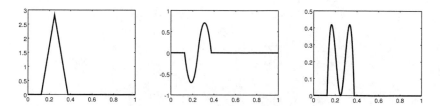

Figure 4.2: Inner quark for $m = 2$ and $p = 0, 1, 2$.

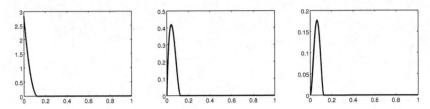

Figure 4.3: First left boundary quark for $m = 3$ and $p = 0, 1, 2$.

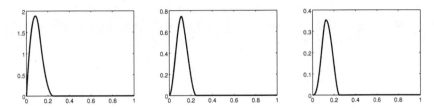

Figure 4.4: Second left boundary quark for $m = 3$ and $p = 0, 1, 2$.

In [28] it was shown that the cardinal B-spline quark is refinable in the following sense.

Proposition 4.3. [28, Prop. 5] For any $p \geq 0$, the vector $(\varphi_0, \ldots, \varphi_p)$ is refinable with $(p + 1) \times (p + 1)$-refinement matrices A_k given by

$$(A_k)_{q,l} := \frac{1}{2^q} a_k \binom{q}{l} k^{q-l}, \qquad (4.1.3)$$

i.e.,

$$\begin{pmatrix} \varphi_0(x) \\ \vdots \\ \varphi_p(x) \end{pmatrix} = \sum_{k \in \mathbb{Z}} A_k \begin{pmatrix} \varphi_0(2x - k) \\ \vdots \\ \varphi_p(2x - k) \end{pmatrix}, \qquad x \in \mathbb{R}. \qquad (4.1.4)$$

Here, the refinement coefficients slightly differ from the ones given in Section 2.1. This is a result of translating the underlying B-spline. For the symmetrised version $\widetilde{N}_m = N_m(\cdot + \lfloor \frac{m}{2} \rfloor)$ the following holds

$$\widetilde{N}_m = \sum_{k=-\lfloor \frac{m}{2} \rfloor}^{\lceil \frac{m}{2} \rceil} 2^{1-m} \binom{m}{k + \lfloor \frac{m}{2} \rfloor} \widetilde{N}_m(2 \cdot -k). \qquad (4.1.5)$$

Analogously to the boundary functions of the Primbs basis, the boundary quarks are not refinable, but fulfil a two-scale relation. This will be shown in the next proposition. Because of the symmetry we restrict our discussion to left boundary quarks.

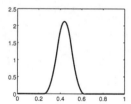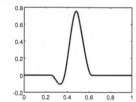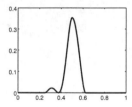

Figure 4.5: Inner quark for $m = 3$ and $p = 0, 1, 2$.

Proposition 4.4. *[77] For every $p \in \mathbb{N}_0$ there exist coefficients $a^j_{q,k,l} \in \mathbb{R}$, so that the left boundary quarks fulfil a two-scale-relation of the form*

$$\varphi_{p,j,k} = \sum_{l=-m+1}^{m-2} \sum_{q=0}^{p} a^j_{q,k,l} \varphi_{q,j+1,l}, \quad k = -m+1, \ldots, -1. \tag{4.1.6}$$

Proof. Let $k \in \{-m+1, \ldots, -1\}$ be fixed. We use the corresponding two-scale-relation for the Primbs wavelet generators, cf. [67, Lem. 3.3]:

$$\varphi_{j,k} = \sum_{l=-m+1}^{m-2} a^j_{k,l} \varphi_{j+1,l}.$$

Inserting this relation into the definition of the left boundary quarks, we obtain

$$\varphi_{p,j,k} = \left(\frac{2^j \cdot}{k+m} \right)^p \varphi_{j,k} = \left(\frac{2^j \cdot}{k+m} \right)^p \sum_{l=-m+1}^{m-2} a^j_{k,l} \varphi_{j+1,l}$$

$$= \left(\frac{2^j \cdot}{k+m} \right)^p \left(\sum_{l=-m+1}^{-1} a^j_{k,l} \varphi_{j+1,l} + \sum_{l=0}^{m-2} a^j_{k,l} \varphi_{j+1,l} \right). \tag{4.1.7}$$

The first sum can be converted into a sum of left boundary quarks of degree p:

$$\left(\frac{2^j \cdot}{k+m} \right)^p \sum_{l=-m+1}^{-1} a^j_{k,l} \varphi_{j+1,l} = \sum_{l=-m+1}^{-1} a^j_{k,l} \left(\frac{l+m}{2(k+m)} \right)^p \frac{(2^{j+1} \cdot)^p}{(l+m)^p} \varphi_{j+1,l}$$

$$= \sum_{l=-m+1}^{-1} a^j_{k,l} \left(\frac{l+m}{2(k+m)} \right)^p \varphi_{p,j+1,l}. \tag{4.1.8}$$

For the second sum we obtain by an application of the binomial theorem

$$(2^j \cdot)^p \sum_{l=0}^{m-2} a^j_{k,l} \varphi_{j+1,l} = 2^{-p} \sum_{l=0}^{m-2} a^j_{k,l} \left(2^{j+1} \cdot -l - \lfloor \tfrac{m}{2} \rfloor + l + \lfloor \tfrac{m}{2} \rfloor \right)^p \varphi_{j+1,l}$$

$$= 2^{-p} \sum_{l=0}^{m-2} a^j_{k,l} \sum_{q=0}^{p} \binom{p}{q} \left(2^{j+1} \cdot -l - \lfloor \tfrac{m}{2} \rfloor \right)^q \left(l + \lfloor \tfrac{m}{2} \rfloor \right)^{p-q} \varphi_{j+1,l}.$$

Putting the monomials and wavelet generators together, we get

$$(2^j \cdot)^p \sum_{l=0}^{m-2} a_{k,l}^j \varphi_{j+1,l} = \sum_{l=0}^{m-2} \sum_{q=0}^{p} a_{k,l}^j 2^{-p} \binom{p}{q} \left(l + \lfloor \tfrac{m}{2} \rfloor \right)^{p-q} \lceil \tfrac{m}{2} \rceil^q$$

$$\cdot \frac{\left(2^{j+1} \cdot -l - \lfloor \tfrac{m}{2} \rfloor\right)^q}{\lceil \tfrac{m}{2} \rceil^q} \varphi_{j+1,l}$$

$$= \sum_{l=0}^{m-2} \sum_{q=0}^{p} a_{k,l}^j 2^{-p} \binom{p}{q} \left(l + \lfloor \tfrac{m}{2} \rfloor \right)^{p-q} \lceil \tfrac{m}{2} \rceil^q \varphi_{q,j+1,l}. \qquad (4.1.9)$$

Combining (4.1.7)-(4.1.9) leads to the following coefficients of the two-scale-relation:

$$a_{q,k,l}^j = \begin{cases} a_{k,l}^j \left(\frac{l+m}{2(k+m)}\right)^p \delta_{p,q}, & l = -m+1, \dots, -1, \\ a_{k,l}^j \left(\frac{1}{2(k+m)}\right)^p \binom{p}{q} \left(l + \lfloor \tfrac{m}{2} \rfloor\right)^{p-q} \lceil \tfrac{m}{2} \rceil^q, & l = 0, \dots, m-2. \end{cases}$$

$\square$

In the sequel it is a great convenience to study the properties of the quarks independently of the level. For this purpose we introduce quarks on level zero on the interval $[0, \infty)$:

$$\varphi_{p,0,k} := \begin{cases} \left(\frac{\cdot}{k+m}\right)^p B_{0,k}^m, & k = -m+1, \dots, -1, \\ \left(\frac{\cdot - k - \lfloor \tfrac{m}{2} \rfloor}{\lceil \tfrac{m}{2} \rceil}\right)^p B_{0,k}^m, & k = 0, 1, \dots, \end{cases} \qquad (4.1.10)$$

where the knot sequence $\{t_k^0\}_{k \geq -m+1}$ is given by

$$t_k^0 := \begin{cases} 0, & k = -m+1, \dots, -1, \\ k, & k = 0, 1, \dots. \end{cases} \qquad (4.1.11)$$

To show the stability of the quarklet systems, bounds for the L_q-norm of the boundary quarks are necessary. In Proposition 4.6 we formulate such a statement. Analogous properties for the inner quarks have been discussed in [28]. We start with an auxiliary result.

Lemma 4.5. *[77] Let $1 \leq k \leq m-1$ and $\varphi_{p,0,-m+k}$ be a left boundary quark. For every $p \geq (m-1)(k-1)$ the unique extremal point of $\varphi_{p,0,-m+k}$ is located at*

$$\hat{x} = \frac{kp}{p+m-1}. \qquad (4.1.12)$$

Proof. Let $x \in \mathbb{R}$. First we have a look at the leftmost quark, i.e., let $k = 1$:

$$\varphi_{p,0,-m+1}(x) = \left(\frac{x}{-m+1+m}\right)^p B_{0,-m+1}^m(x) = x^p B_{0,-m+1}^m(x).$$

By using the differentiation rules and the recursive form of the B-splines, cf. [75, Thm. 4.15, 4.16], we obtain

$$
\begin{aligned}
\varphi'_{p,0,-m+1}(x) &= px^{p-1}B^m_{0,-m+1}(x) + x^p(B^m_{0,-m+1})'(x) \\
&= px^{p-1}B^m_{0,-m+1}(x) - x^p(m-1)B^{m-1}_{0,-m+2}(x) \\
&= px^{p-1}\frac{t_1 - x}{t_1 - t_{-m+2}}B^{m-1}_{0,-m+2}(x) - x^p(m-1)B^{m-1}_{0,-m+2}(x) \\
&= x^{p-1}\left(p(1-x) - x(m-1)\right)B^{m-1}_{0,-m+2}(x).
\end{aligned}
$$

We obtain the critical points $x = 0$, where the B-spline and also the quark is zero, and $\hat{x} = \frac{p}{p+m-1}$, where $|\varphi_{p,0,-m+1}|$ attends its maximum. Now assume $m \geq 3$, $k \geq 2$ and let $\varphi_{p,0,-m+k}$ be the k-th left boundary quark:

$$
\varphi_{p,0,-m+k}(x) = \left(\frac{x}{-m+k+m}\right)^p B^m_{0,-m+k}(x) = k^{-p}x^p B^m_{0,-m+k}(x).
$$

The support of the quark $\varphi_{p,0,-m+k}$ is the interval $[0,k]$. In the first step we show that $\varphi_{p,0,-m+k}$ is monotonically increasing on $[0, k-1]$. For the first derivative we estimate

$$
\begin{aligned}
\varphi'_{p,0,-m+k}(x) &= k^{-p}px^{p-1}B^m_{0,-m+k}(x) + k^{-p}x^p(B^m_{0,-m+k})'(x) \\
&= k^{-p}x^{p-1}\left(pB^m_{0,-m+k}(x) + x(B^m_{0,-m+k})'(x)\right) \\
&\geq k^{-p}x^{p-1}\left(pB^m_{0,-m+k}(x) - \left|x(B^m_{0,-m+k})'(x)\right|\right).
\end{aligned}
$$

Again, we use the differentiation rules and recursion to derive

$$
\begin{aligned}
\varphi'_{p,0,-m+k}(x) &\geq k^{-p}x^{p-1}\left(pB^m_{0,-m+k}(x) - \left|x(m-1)\left(\frac{B^{m-1}_{0,-m+k}(x)}{k-1} - \frac{B^{m-1}_{0,-m+k+1}(x)}{k}\right)\right|\right) \\
&\geq k^{-p}x^{p-1}\left(pB^m_{0,-m+k}(x) - x(m-1)\left(\frac{B^{m-1}_{0,-m+k}(x)}{k-1} + \frac{B^{m-1}_{0,-m+k+1}(x)}{k}\right)\right).
\end{aligned}
$$

For $x \in [0,1]$ it holds $k - x \geq x$, which yields

$$
\begin{aligned}
\varphi'_{p,0,-m+k}(x) &\geq k^{-p}x^{p-1} \\
&\quad \cdot \left(pB^m_{0,-m+k}(x) - (m-1)\left(\frac{x}{k-1}B^{m-1}_{0,-m+k}(x) + \frac{k-x}{k}B^{m-1}_{0,-m+k+1}(x)\right)\right) \\
&= k^{-p}x^{p-1}\left(pB^m_{0,-m+k}(x) - (m-1)B^m_{0,-m+k}(x)\right) \\
&= k^{-p}x^{p-1}\left(p - (m-1)\right)B^m_{0,-m+k}(x).
\end{aligned}
$$

Hence, the derivative is non-negative on $[0,1]$ if $p \geq m-1$. For $x \in [1, k-1]$, it trivially holds $x \geq 1$ and $k - x \geq 1$. It follows

$$\varphi'_{p,0,-m+k}(x) \geq k^{-p}x^{p-1}\left(pB^m_{0,-m+k}(x) - x\left|(B^m_{0,-m+k})'(x)\right|\right)$$

$$\geq k^{-p}x^{p-1}\left(pB^m_{0,-m+k}(x) - (k-1)\left|(B^m_{0,-m+k})'(x)\right|\right)$$

$$= k^{-p}x^{p-1}$$

$$\cdot \left(pB^m_{0,-m+k}(x) - (k-1)\left|(m-1)\left(\frac{B^{m-1}_{0,-m+k}(x)}{k-1} - \frac{B^{m-1}_{0,-m+k+1}(x)}{k}\right)\right|\right).$$

By the above considerations we can further estimate

$$\varphi'_{p,0,-m+k}(x)$$

$$\geq k^{-p}x^{p-1}\left(pB^m_{0,-m+k}(x) - (k-1)(m-1)\left(\frac{1}{k-1}B^{m-1}_{0,-m+k}(x) + \frac{1}{k}B^{m-1}_{0,-m+k+1}(x)\right)\right)$$

$$\geq k^{-p}x^{p-1}$$

$$\cdot \left(pB^m_{0,-m+k}(x) - (k-1)(m-1)\left(\frac{x}{k-1}B^{m-1}_{0,-m+k}(x) + \frac{k-x}{k}B^{m-1}_{0,-m+k+1}(x)\right)\right).$$

With the recursive relation of the B-splines we get

$$\varphi'_{p,0,-m+k}(x) \geq k^{-p}x^{p-1}\left(pB^m_{0,-m+k}(x) - (k-1)(m-1)B^m_{0,-m+k}(x)\right)$$

$$= k^{-p}x^{p-1}\left(p - (k-1)(m-1)\right)B^m_{0,-m+k}(x).$$

Finally we conclude that for $p \geq (m-1)(k-1)$ the derivative is non-negative on $[1, k-1]$. So all extremal points are located in $[k-1, k]$, where we can compute an explicit form of $\varphi_{p,0,-m+k}$. To do this, we first compute the explicit form of $B^m_{0,-m+k}$. From the definition of the B-splines and the recursion for divided differences we get:

$$B^m_{0,-m+k}(x) = (t^0_k - t^0_{-m+k})(\cdot - x)^{m-1}_+[t^0_{-m+k}, \ldots, t^0_k]$$

$$= \frac{k}{k}\left((\cdot - x)^{m-1}_+[t^0_{-m+k+1}, \ldots, t^0_k] - (\cdot - x)^{m-1}_+[t^0_{-m+k}, \ldots, t^0_{k-1}]\right)$$

$$= (\cdot - x)^{m-1}_+[t^0_{-m+k+1}, \ldots, t^0_k].$$

The latter divided difference vanishes, because of $x \geq k-1$. On the interval $[0, k-1]$ the truncated polynomial $(\cdot - x)^{m-1}_+$ is zero. Hence all of the coefficients of the interpolating polynomial are zero. By repeating this argument $m - k - 1$ times we obtain

$$B^m_{0,-m+k}(x) = k^{-1-(m-k-1)}(\cdot - x)^{m-1}_+[1, \ldots, k].$$

After another $k - 1$ iterations we obtain

$$B^m_{0,-m+k}(x) = k^{-m+k}\frac{1}{(k-1)!}(\cdot - x)^{m-1}_+[k].$$

We end up with

$$B^m_{0,-m+k}|_{[k-1,k]}(x) = k^{-m+k}\frac{1}{(k-1)!}(k-x)^{m-1}.$$

With this representation we compute the derivative $\varphi'_{p,0,-m+k}$ on $[k-1,k]$:

$$\varphi'_{p,0,-m+k}(x) = k^{-p}px^{p-1}B^m_{0,-m+k}(x) + k^{-p}x^p B^{m\prime}_{0,-m+k}(x)$$
$$= k^{-p}x^{p-1}$$
$$\cdot\left(pk^{-m+k}\frac{1}{(k-1)!}(k-x)^{m-1} - xk^{-m+k}\frac{1}{(k-1)!}(m-1)(k-x)^{m-2}\right)$$
$$= k^{-p-m+k}x^{p-1}\frac{1}{(k-1)!}\left((k-x)^{m-2}(p(k-x) - x(m-1))\right).$$

We obtain the critical points $x = 0$, $x = k$, where $B^m_{0,-m+k}$ is zero, and $\hat{x} = \frac{kp}{p+m-1}$, where $|\varphi_{p,0,-m+k}|$ attains its maximum. Indeed $\hat{x}$ lies in $[k-1,k]$, because on the one hand we have

$$\hat{x} = \frac{kp}{p+m-1} \leq \frac{kp+k(m-1)}{p+m-1} = \frac{k(p+m-1)}{p+m-1} = k.$$

On the other hand it holds true that

$$k-1 = k-\frac{k(m-1)}{k(m-1)} = k-\frac{k(m-1)}{(m-1)(k-1)+m-1} \leq k-\frac{k(m-1)}{p+m-1} = \frac{kp}{p+m-1} = \hat{x}.$$

$\square$

Proposition 4.6. *[77] Let $1 \leq k \leq m-1$ and $\varphi_{p,0,-m+k}$ be a left boundary quark. For every $1 \leq q \leq \infty$ there exist constants $c = c(m,k,q) > 0$, $C = C(m,k,q) > 0$, so that for all $p \geq (m-1)(k-1)$ it holds:*

$$c(p+1)^{-(m-1+1/q)} \leq \|\varphi_{p,0,-m+k}\|_{L_q(\mathbb{R})} \leq C(p+1)^{-(m-1+1/q)}. \qquad (4.1.13)$$

Proof. We show (4.1.13) for the extremal cases $q \in \{1,\infty\}$ and conclude by Hölder's inequality. To derive the upper bound for $q = 1$ we use an integration formula for general B-splines and functions $f \in C^m([t^0_{-m+k}, t^0_k])$, cf. [75, Thm. 4.23]:

$$\int_{t^0_{-m+k}}^{t^0_k} B^m_{0,-m+k}(x)f^{(m)}(x)\,dx = (t^0_k - t^0_{-m+k})(m-1)!f[t^0_{-m+k},\ldots,t^0_k].$$

Choosing $f(x) := x^{p+m}\frac{1}{(p+m)\cdots(p+1)}$ we obtain

$$\|\varphi_{p,0,-m+k}\|_{L_1(\mathbb{R})} = \left(\frac{1}{k}\right)^p \int_{t^0_{-m+k}}^{t^0_k} B^m_{0,-m+k}(x)x^p\,dx$$
$$= \left(\frac{1}{k}\right)^p (k-0)(m-1)!\,(\cdot)^{p+m}[t^0_{-m+k},\ldots,t^0_k]\frac{1}{(p+m)\cdots(p+1)}$$
$$\leq \left(\frac{1}{k}\right)^{p-1}(m-1)!\,(\cdot)^{p+m}[t^0_{-m+k},\ldots,t^0_k](p+1)^{-m}.$$

To estimate the divided difference we use a Leibniz rule with $x^{p+m} = xx^{p+m-1}$, cf. [75, Thm. 2.52]:

$$(\cdot)^{p+m}[t^0_{-m+k}, \dots, t^0_k] = \sum_{i=-m+k}^{k} (\cdot)^1[t^0_{-m+k}, \dots, t^0_i] \, (\cdot)^{p+m-1}[t^0_i, \dots, t^0_k].$$

For the first order polynomial all but one non-trivial summand can be eliminated:

$$\begin{aligned}
(\cdot)^{p+m}[t^0_{-m+k}, \dots, t^0_k] &= (\cdot)^1[t^0_{-m+k}] \, (\cdot)^{p+m-1}[t^0_{-m+k}, \dots, t^0_k] \\
&\quad + (\cdot)^1[t^0_{-m+k}, t^0_{-m+k+1}] \, (\cdot)^{p+m-1}[t^0_{-m+k+1}, \dots, t^0_k] \\
&= (\cdot)^{p+m-1}[t^0_{-m+k+1}, \dots, t^0_k].
\end{aligned}$$

Repeating this argument $m-k$ times we get

$$(\cdot)^{p+m}[t_{-m+k}, \dots, t_k] = (\cdot)^{p+k}[t^0_0, \dots, t^0_k].$$

By eliminating the leading zeros we get equidistant knots and can replace the divided difference by a forward difference, cf. [75, Thm. 2.57]:

$$\begin{aligned}
(\cdot)^{p+m}[t^0_{-m+k}, \dots, t^0_k] &= \frac{1}{k!}(\Delta^k(\cdot)^{p+k})(0) \\
&= \frac{1}{k!}\sum_{j=0}^{k}\binom{k}{j}(-1)^{k-j}j^{p+k} \\
&\leq \frac{1}{k!}k^p\sum_{j=0}^{k}\binom{k}{j}j^k.
\end{aligned}$$

Finally we get the upper estimate with $C(m,k) = \frac{(m-1)!}{(k-1)!}\sum_{j=0}^{k}\binom{k}{j}j^k$:

$$\|\varphi_{p,0,-m+k}\|_{L_1(\mathbb{R})} \leq C(p+1)^{-m}. \tag{4.1.14}$$

Now let $q = \infty$. With (4.1.12) we directly compute

$$\begin{aligned}
\|\varphi_{p,0,-m+k}\|_{L_\infty(\mathbb{R})} = |\varphi_{p,0,-m+k}(\hat{x})| &= k^{-p}\hat{x}^p k^{-m+k}\frac{1}{(k-1)!}(k-\hat{x})^{m-1} \\
&= \frac{k^{-m+k}}{(k-1)!}\left(\frac{p}{p+m-1}\right)^p\left(\frac{k(m-1)}{p+m-1}\right)^{m-1}.
\end{aligned}$$

We get the upper estimate with some constant $C(m,k) = \frac{k^{-m+k}}{(k-1)!}(k(m-1))^{m-1}$:

$$\|\varphi_{p,0,-m+k}\|_{L_\infty(\mathbb{R})} \leq C(p+1)^{-(m-1)}. \tag{4.1.15}$$

For $1 < q < \infty$ an application of Hölder's inequality and (4.1.14), (4.1.15) yields

$$\begin{aligned}
\|\varphi_{p,0,-m+k}\|_{L_q(\mathbb{R})}^q &\leq \|\varphi_{p,0,-m+k}\|_{L_1(\mathbb{R})}^{1/q}\|\varphi_{p,0,-m+k}\|_{L_\infty(\mathbb{R})}^{1-1/q} \\
&\leq C(p+1)^{-(m-1-1/q)},
\end{aligned}$$

which proves the upper estimate. Now we turn to the lower estimate. Let $q = \infty$. From our previous calculations we directly get the lower estimate with $c(m,k) = \tilde{c}e^{1-m}\frac{k^{-m+k}}{(k-1)!}(k(m-1))^{m-1}$, where $\tilde{c} > 0$ just depends on m:

$$c(p+1)^{-(m-1)} \le \|\varphi_{p,0,-m+k}\|_{L_\infty(\mathbb{R})}. \tag{4.1.16}$$

It remains to show the lower estimate for $q \in \mathbb{N}$. An elementary estimate leads to

$$\|\varphi_{p,0,-m+k}\|^q_{L_q(\mathbb{R})} = \int_0^k |\varphi_{p,0,-m+k}(x)|^q \, dx$$

$$\ge \int_{k-1}^k |\varphi_{p,0,-m+k}(x)|^q \, dx$$

$$= \int_{k-1}^k \left(\frac{x}{k}\right)^{pq} \left(B^m_{0,-m+k}(x)\right)^q \, dx$$

$$= \int_{k-1}^k \left(\frac{x}{k}\right)^{pq} \left(k^{-m+k}\frac{1}{(k-1)!}(k-x)^{m-1}\right)^q \, dx.$$

Substitution leads to

$$\|\varphi_{p,0,-m+k}\|^q_{L_q(\mathbb{R})} \ge \frac{1}{((k-1)!)^q}k^{(-m+k)q}\int_{k-1}^k \left(\frac{x}{k}\right)^{pq}(k-x)^{(m-1)q} \, dx$$

$$= \frac{1}{((k-1)!)^q}k^{(-m+k)q}\int_0^1 \left(\frac{k-y}{k}\right)^{pq}y^{(m-1)q} \, dy$$

$$\ge \frac{1}{((k-1)!)^q}k^{(-m+k)q}\int_0^1 (1-y)^{pq}y^{(m-1)q} \, dy.$$

By $m(q-1)$ times partial integration we obtain

$$\int_0^1 (1-y)^{pq}y^{(m-1)q} \, dy = \frac{(m-1)q}{pq+1}\int_0^1 (1-y)^{pq+1}y^{(m-1)q-1} \, dy = \cdots$$

$$= \frac{((m-1)q)!}{(pq+1)(pq+2)\cdots(pq+mq-q)}\frac{1}{pq+mq-q+1}.$$

We go on estimating by

$$\int_0^1 (1-y)^{pq}y^{(m-1)q} \, dy \ge \frac{((m-1)q)!}{(\tilde{c}(p+1))^{mq-q+1}},$$

where $\widetilde{c} > 0$ just depends on m and q. Finally we get the lower estimate with
$c(m, k, q) = \frac{k^{-m+k}}{(k-1)!} \left(\frac{((m-1)q)!}{\widetilde{c}} \right)^{1/q}$:

$$\|\varphi_{p,0,-m+k}\|_{L_q(\mathbb{R})} \geq c(p+1)^{-(m-1+1/q)}. \tag{4.1.17}$$

For $1 < q < \infty$ we again use Hölder's inequality. First let $1 < q \leq 2$, then by
(4.1.16),(4.1.17) it follows

$$\|\varphi_{p,0,-m+k}\|_{L_q(\mathbb{R})}^q \geq \|\varphi_{p,0,-m+k}\|_{L_2(\mathbb{R})}^{2/q} \|\varphi_{p,0,-m+k}\|_{L_\infty(\mathbb{R})}^{1-2/q}$$
$$\geq c(p+1)^{-(m-1-1/q)}.$$

For $2 \leq q < \infty$, using (4.1.17) we have

$$\|\varphi_{p,0,-m+k}\|_{L_q(\mathbb{R})}^q \geq \|\varphi_{p,0,-m+k}\|_{L_2(\mathbb{R})}^{2-2/q} \|\varphi_{p,0,-m+k}\|_{L_1(\mathbb{R})}^{2/q-1}$$
$$\geq c(p+1)^{-(m-1-1/q)},$$

which completes the proof. $\qquad\qquad\qquad\qquad\qquad\qquad\qquad\qquad\square$

In [28], an analogous statement for cardinal B-spline quarks has been proven.

Proposition 4.7. *[28, Prop. 1] Let φ_p as in (4.1.1). For every $1 \leq q \leq \infty$ there
exist constants $c = c(m, k, q) > 0$, $C = C(m, k, q) > 0$, so that for all $p \geq (m-1)^2$:*

$$c(p+1)^{-(m-1+1/q)} \leq \|\varphi_p\|_{L_q(\mathbb{R})} \leq C(p+1)^{-(m-1+1/q)}. \tag{4.1.18}$$

In the following we study regularity properties of the quarks. Our aim is to derive
Bernstein inequalities in Besov spaces. We rewrite (3.2) as

$$V_{p,j} := \overline{\text{span}\{\varphi_q(2^j \cdot -k) : 0 \leq q \leq p, k \in \mathbb{Z}\}}^{L_q}, \tag{4.1.19}$$

The following two Bernstein inequalities have been proven in [28]. They are formulated for the case of the real line, but are valid for the boundary adapted case,
too.

Theorem 4.8. *[28, Thm. 2] Let the modulus of smoothness be defined as in (1.1.24).
Then, for $1 \leq q \leq \infty$, there exists $C(m, q)$ such that for all $f \in V_{p,j}$ it holds*

$$\omega_m(f, t)_{L_q(\mathbb{R})} \leq C \min\{1, (p+1)^2 2^j t\}^{m-1+1/q} \|f\|_{L_q(\mathbb{R})}. \tag{4.1.20}$$

Corollary 4.9. *[28, Cor. 1] For $1 \leq q \leq \infty$, there exists $C(m, q)$ such that for all
$f \in V_{p,j}$ it holds*

$$\|f^{(k)}\|_{L_q(\mathbb{R})} \leq C(p+1)^{2k} 2^{jk} \|f\|_{L_q(\mathbb{R})}. \tag{4.1.21}$$

We use the first result to prove an obvious generalisation of [28, Cor. 2]. Note
that just the constant in (4.1.22) depends on the fine tuning parameter of the Besov
spaces.

Theorem 4.10. *For all $0 < s < m - 1 + \frac{1}{q}$, $1 \le r, q < \infty$ there exists $C(m, q, r, s)$ such that for all $f \in V_{p,j}$ it holds*

$$|f|_{B^s_r(L_q(\mathbb{R}))} \le C(p+1)^{2s} 2^{js} \|f\|_{L_q(\mathbb{R})}. \tag{4.1.22}$$

Proof. We use the integral semi-norm from (1.1.29) for $s < m$. Splitting the range of integration leads to

$$|f|^r_{B^s_r(L_q(\mathbb{R}))} = \int_0^\infty \left(t^{-s}\omega_m(f,t)_{L_q(\mathbb{R})}\right)^r \frac{dt}{t}$$

$$= \int_0^{(p+1)^{-2}2^{-j}} \left(t^{-s}\omega_m(f,t)_{L_q(\mathbb{R})}\right)^r \frac{dt}{t} + \int_{(p+1)^{-2}2^{-j}}^\infty \left(t^{-s}\omega_m(f,t)_{L_q(\mathbb{R})}\right)^r \frac{dt}{t}.$$

With (4.1.20) we obtain

$$|f|^r_{B^s_r(L_q(\mathbb{R}))} \lesssim \int_0^{(p+1)^{-2}2^{-j}} \left(t^{-s}((p+1)^2 2^j t)^{m-1+1/q}\|f\|_{L_q(\mathbb{R})}\right)^r \frac{dt}{t}$$

$$+ \int_{(p+1)^{-2}2^{-j}}^\infty \left(t^{-s}\|f\|_{L_q(\mathbb{R})}\right)^r \frac{dt}{t}.$$

Explicitly calculating the integrals yields

$$|f|^r_{B^s_r(L_q(\mathbb{R}))} \lesssim \|f\|^r_{L_q(\mathbb{R})}((p+1)^2 2^j)^{r(m-1+1/q)} \int_0^{(p+1)^{-2}2^{-j}} t^{r(-s+m-1+1/q)-1}\, dt$$

$$+ \|f\|^r_{L_q(\mathbb{R})} \int_{(p+1)^{-2}2^{-j}}^\infty t^{-rs-1}\, dt$$

$$= \|f\|^r_{L_q(\mathbb{R})}(p+1)^{2rs}2^{jrs}\left(\frac{1}{r(m-1+\frac{1}{q}-s)} - \frac{1}{rs}\right).$$

For the above calculations we used that the condition $0 < s < m-1+\frac{1}{q}$ ensured the existence of the integrals. Hence, the claim follows. $\square$

We transfer these findings to the case of $q < 1$.

Theorem 4.11. *Let the modulus of smoothness be defined as in (1.1.24). Then, for $0 < q < 1$, there exists $C(m,q)$ such that for all $f \in V_{p,j}$ it holds*

$$\omega_m(f,t)_{L_q(\mathbb{R})} \le C \min\{1, (p+1)^{2m/q}(2^j t)^{m-1+1/q}\}\|f\|_{L_q(\mathbb{R})}. \tag{4.1.23}$$

Proof. Let $t > (p+1)^{-\frac{m}{q(m+1-\frac{1}{q})}} 2^{-j}$. With the definition (1.1.24) and the inequality $\|f+g\|_{L_q}^q \le \|f\|_{L_q}^q + \|g\|_{L_q}^q$ we obtain

$$\omega_m(f,t)_{L_q(\mathbb{R})}^q := \sup_{|h|\le t} \|\Delta_h^m f\|_{L_q(\mathbb{R})}^q$$

$$\le \sup_{|h|\le t} \|\Delta_h^{m-1} f(\cdot + h)\|_{L_q(\mathbb{R})}^q + \|\Delta_h^{m-1} f\|_{L_q(\mathbb{R})}^q$$

$$\le \sup_{|h|\le t} 2\|\Delta_h^{m-1} f\|_{L_q(\mathbb{R})}^q$$

Inductively we get

$$\omega_m(f,t)_{L_q(\mathbb{R})}^q \le 2^m \|f\|_{L_q(\mathbb{R})}^q.$$

Hence, for this case (4.1.23) holds with a constant $C = 2^{m/q}$. Now let $t \le (p+1)^{-\frac{m}{q(m+1-\frac{1}{q})}} 2^{-j}$. Since the quarks are differentiable almost everywhere, we can calculate

$$\|\Delta_h^m f\|_{L_q(\mathbb{R})}^q = \|\Delta_h^{m-1} f(\cdot + h) - \Delta_h^{m-1} f\|_{L_q(\mathbb{R})}^q$$

$$= \int_{\mathbb{R}} \left| \int_x^{x+h} \Delta_h^{m-1} f'(y) \, \mathrm{d}y \right|^q \mathrm{d}x$$

$$\le \int_{\mathbb{R}} \|\Delta_h^{m-1} f'\|_{L_1(x,x+h)}^q \, \mathrm{d}x.$$

With a Nikolski type inequality, cf. [83, Sec. 4.9.6], we obtain

$$\|\Delta_h^m f\|_{L_q(\mathbb{R})}^q \le h^{(1-\frac{1}{q})q}(p+m-2)^{2(\frac{1}{q}-1)q} \int_{\mathbb{R}} \|\Delta_h^{m-1} f'\|_{L_q(x,x+h)}^q \, \mathrm{d}x$$

$$= h^{q-1}(p+m-2)^{2(1-q)} h \|\Delta_h^{m-1} f'\|_{L_q(\mathbb{R})}^q.$$

Inductively we obtain with a constant $C_1 > 0$ depending on m, q:

$$\|\Delta_h^m f\|_{L_q(\mathbb{R})}^q \le C_1 h^{q(m-1)}(p+1)^{2(1-q)(m-1)} \|\Delta_h f^{(m-1)}\|_{L_q(\mathbb{R})}^q. \qquad (4.1.24)$$

Splitting the range of integration and applying the mean value theorem yields

$$\|\Delta_h f^{(m-1)}\|_{L_q(\mathbb{R})}^q = \sum_{l\in\mathbb{Z}} \|f^{(m-1)}(\cdot + h) - f^{(m-1)}\|_{L_q(2^{-j}l, 2^{-j}(l+1))}^q$$

$$\le \sum_{l\in\mathbb{Z}} \left(h^q \|f^{(m)}\|_{L_q(2^{-j}l, 2^{-j}(l+1))}^q + \|f^{(m-1)}\|_{L_q(2^{-j}l, 2^{-j}l+h)}^q \right.$$

$$\left. + \|f^{(m-1)}\|_{L_q(2^{-j}(l+1)-h, 2^{-j}(l+1))}^q \right).$$

An application of standard inequalities for polynomials, see again [83, Sec. 4.9.6] leads to

$$\|f^{(m-1)}\|^q_{L_q(2^{-j}l,2^{-j}l+h)} + \|f^{(m-1)}\|^q_{L_q(2^{-j}(l+1)-h,2^{-j}(l+1))}$$
$$\lesssim h2^j(p+1)^2\|f^{(m-1)}\|^q_{L_q(2^{-j}l,2^{-j}(l+1))}.$$

With a Markov inequality we get

$$\|\Delta_h^m f\|^q_{L_q(\mathbb{R})} \lesssim \left(h^q(p+1)^{2q}2^{jq} + h2^j(p+1)^2\right)\|f^{(m-1)}\|^q_{L_q(\mathbb{R})}.$$

Applying the Markov inequality inductively $m-1$ times, we obtain

$$\|\Delta_h^m f\|^q_{L_q(\mathbb{R})} \lesssim \left(h^q(p+1)^{2q}2^{jq} + h2^j(p+1)^2\right)\frac{((p+1)\cdots(p+m-1))^{2q}}{2^{-jq(m-1)}}\|f\|^q_{L_q(\mathbb{R})}$$
$$\lesssim \left(h^q(p+1)^{2q}2^{jq} + h2^j(p+1)^2\right)(p+1)^{2q(m-1)}2^{jq(m-1)}\|f\|^q_{L_q(\mathbb{R})}.$$

Putting all together and estimating $h < 2^{-j}$ gives

$$\|\Delta_h^m f\|^q_{L_q(\mathbb{R})} \lesssim (p+1)^{2m}h^{q(m-1)+1}2^{jq(m-1)+j}\|f\|^q_{L_q(\mathbb{R})},$$

and hence the claim follows. $\qquad\square$

Theorem 4.12. *For all $\frac{1}{q} - 1 < s < m$, $0 < r,q < 1$ there exists $C(m,q,r,s)$ such that for all $f \in V_{p,j}$ it holds*

$$|f|_{B_r^s(L_q(\mathbb{R}))} \leq C(p+1)^{2s\frac{m}{(m-1+1/q)q}}2^{js}\|f\|_{L_q(\mathbb{R})}. \qquad (4.1.25)$$

Proof. The proof is analogous to the proof of Theorem 4.10. We use the integral quasi-semi-norm from (1.1.28) for the specified range of s. Splitting the range of integration leads to

$$|f|^r_{B_r^s(L_q(\mathbb{R}))} = \int_0^\infty \left(t^{-s}\omega_m(f,t)_{L_q(\mathbb{R})}\right)^r \frac{dt}{t}$$

$$= \int_0^{(p+1)^{-\frac{2m}{q(m-1+1/q)}}2^{-j}} \left(t^{-s}\omega_m(f,t)_{L_q(\mathbb{R})}\right)^r \frac{dt}{t}$$

$$+ \int_{(p+1)^{-\frac{2m}{q(m-1+1/q)}}2^{-j}}^\infty \left(t^{-s}\omega_m(f,t)_{L_q(\mathbb{R})}\right)^r \frac{dt}{t}.$$

With (4.1.23) we obtain

$$|f|^r_{B_r^s(L_q(\mathbb{R}))} \lesssim \int_0^{(p+1)^{-\frac{2m}{q(m-1+1/q)}}2^{-j}} \left(t^{-s}(p+1)^{2m/q}(2^jt)^{m-1+1/q}\|f\|_{L_q(\mathbb{R})}\right)^r \frac{dt}{t}$$

$$+ \int_{(p+1)^{-\frac{2m}{q(m-1+1/q)}}2^{-j}}^\infty \left(t^{-s}\|f\|_{L_q(\mathbb{R})}\right)^r \frac{dt}{t}.$$

Explicitly calculating the integrals yields

$$
|f|^r_{B^s_r(L_q(\mathbb{R}))} \lesssim \|f\|^r_{L_q(\mathbb{R})} (p+1)^{\frac{2mr}{q}} 2^{jr(m-1+\frac{1}{q})} \int\limits_0^{(p+1)^{-\frac{2m}{q(m-1+\frac{1}{q})}} 2^{-j}} t^{r(-s+m-1+\frac{1}{q})-1}\, dt
$$

$$
+ \|f\|^r_{L_q(\mathbb{R})} \int\limits_{(p+1)^{-\frac{2m}{q(m-1+\frac{1}{q})}} 2^{-j}}^{\infty} t^{-rs-1}\, dt
$$

$$
= \|f\|^r_{L_q(\mathbb{R})} (p+1)^{\frac{2mrs}{q(m-1+1/q)}} 2^{jrs} \left(\frac{1}{r(m-1+\frac{1}{q}-s)} - \frac{1}{rs} \right).
$$

Again, we used that the condition $\frac{1}{q} - 1 < s < m$ ensures the existence of the above integrals. $\qquad\square$

4.2 Quarklets on the Real Line

In this section we introduce quarklets in the shift-invariant setting. Analogously to wavelets, quarklets are linear combinations of quark generators. Since multiquarks are refinable function vectors, vectors of quarklets fit in the setting of multiwavelets. Accordingly, we can apply our findings from Chapter 3 to construct quarklets that fulfil certain reconstruction properties. It turns out that, by keeping the wavelet mask for the definition of quarklets, these reconstruction properties hold. This result is presented in Theorem 4.15. In addition we provide an estimation for the complexity of the reconstruction sequences in Theorem 4.18 and give some examples. We start with the definition of cardinal B-spline quarklets on the real line.

Definition 4.13. Let $\varphi = N_m(\cdot + \lfloor \frac{m}{2} \rfloor)$ denote the symmetrised cardinal B-spline with symbol $a(z)$ and dual generator $\tilde{\varphi}$. Let ψ denote the CDF wavelet defined by (2.1.25), (2.1.29). Let the quark φ_p be defined as in (4.1.1). Then, the p-th *quarklet* $\psi_p, p \in \mathbb{N}_0$ is defined by

$$
\psi_p := \sum_{k \in \mathbb{Z}} b_k \varphi_p(2 \cdot - k). \tag{4.2.1}
$$

Additionally we consider scaled, dilated and translated versions of the mother quarklet ψ_p defined in (4.2.1):

$$
\psi_{p,j,k} := 2^{j/2} \psi_p(2^j \cdot - k), \quad p \in \mathbb{N}_0, j \in \mathbb{Z}, k \in \mathbb{Z}. \tag{4.2.2}
$$

In other words, we construct a quarklet by keeping the wavelet mask $b \in \ell_1(\mathbb{Z})$ and using the enriched generators in the corresponding two-scale-equation. Similarly to Section 2.1, it immediately follows that the quarklets fulfil the relation

$$
\hat{\psi}_p(\xi) = \frac{1}{2} b(z) \hat{\varphi}_p \left(\frac{\xi}{2} \right), \quad z = e^{-i\frac{\xi}{2}}, \xi \in \mathbb{R}. \tag{4.2.3}
$$

In Section 4.1 we have seen that the spaces $V_{p,j}$ as defined in (4.1.19) provide a generalised multiresolution structure. Having in mind the wavelet construction as in Section 2.1, the question occurs whether the quarklets ψ_p generate complement spaces $W_{p,j}$. To answer this question, we employ the technique derived in Chapter 3. From Proposition 4.3 we already know that multiquarks with a fixed degree p are refinable. Rearranging the quarks in a vector $\Phi := (\varphi_0, \ldots, \varphi_p)^T$ leads to

$$\hat{\Phi}(\xi) = \frac{1}{2}\mathscr{A}(z)\hat{\Phi}\left(\frac{\xi}{2}\right), \quad \xi \in \mathbb{R}, z = e^{-i\frac{\xi}{2}} \in S_1, \tag{4.2.4}$$

where

$$(\mathscr{A}(z))_{q,l} := \sum_{k \in \mathbb{Z}} (A_k)_{q,l} z^k, \tag{4.2.5}$$

with matrices A_k as in (4.1.3). Note that the matrix indexing starts at zero. Analogously rearranging the quarklets in a vector $\Psi := (\psi_0, \ldots, \psi_p)^T$, we obtain

$$\hat{\Psi}(\xi) = \frac{1}{2}b(z)I\hat{\Phi}\left(\frac{\xi}{2}\right). \tag{4.2.6}$$

Accordingly, the abstract setting from Chapter 3 simplifies to the case of lower triangular matrices and diagonal matrices, respectively. It remains to consider the invertibility of the matrix $X(z)$ in (3.13). Let us recall an important fact concerning the construction of biorthogonal wavelets. Let $a(z), \tilde{a}(z)$ be the symbols of the primal (dual) generator. The biorthogonality of the generators (2.1.18) and the definition of the wavelet symbols (2.1.29) imply the fundamental identity

$$a(z)b(-z) - b(z)a(-z) = 4z. \tag{4.2.7}$$

Proposition 4.14. *Let $z \in S_1$. With the definition*

$$X(z) := \frac{1}{2}\begin{pmatrix} \mathscr{A}(z) & \mathscr{A}(-z) \\ b(z)I & b(-z)I \end{pmatrix}, \tag{4.2.8}$$

it holds that

$$\det X(z) = 2^{-p(p+1)/2}z^{p+1}. \tag{4.2.9}$$

Hence $X(z)$ is invertible on S_1. Furthermore, $X(z)^{-1}$ is given by

$$X(z)^{-1} = 2\begin{pmatrix} b(-z)T(z)^{-1} & -T(z)^{-1}\mathscr{A}(-z) \\ -b(z)T(z)^{-1} & T(z)^{-1}\mathscr{A}(z) \end{pmatrix}, \tag{4.2.10}$$

where

$$T(z) = \mathscr{A}(z)b(-z) - b(z)\mathscr{A}(-z). \tag{4.2.11}$$

Proof. The matrix $T(z) = \mathscr{A}(z)b(-z) - b(z)\mathscr{A}(-z)$ is of lower triangular shape with diagonal entries

$$a_{qq}(z)b(-z) - b(z)a_{qq}(-z) = 2^{-q}(a(z)b(-z) - b(z)a(-z))$$
$$= 2^{-q}4z,$$

where we used (3.6), (4.1.3), (4.2.7). Hence, we conclude

$$\det T(z) = \prod_{q=0}^{p} 2^{-q}4z$$
$$= (4z)^{p+1}2^{-\sum_{q=0}^{p}q}$$
$$= z^{p+1}2^{2p+2-p(p+1)/2}.$$

This implies that $T(z)$ is invertible on S_1. Since $b(z)I$ and $b(-z)I$ commute, the determinant of $X(z)$ can by computed by, cf. [78],

$$\det X(z) = \left(\frac{1}{2}\right)^{2p+2}\det(\mathscr{A}(z)b(-z)I - b(z)I\mathscr{A}(-z))$$
$$= \left(\frac{1}{2}\right)^{2p+2}\det T(z)$$
$$= z^{p+1}2^{-p(p+1)/2}.$$

Finally, one easily verifies (4.2.10). $\qquad\square$

We formulate the first main result of this section.

Theorem 4.15. *Let $\rho \in \{0,1\}$. There exist matrices C, D such that each vector of fine quarks has a decomposition into quarklets and quarks on a coarse level:*

$$\Phi(2 \cdot -\rho) = \sum_{k\in\mathbb{Z}} C_{\rho+2k}\Phi(\cdot - k) + \sum_{n\in\mathbb{Z}} D_{\rho+2n}\Psi(\cdot - n). \qquad (4.2.12)$$

Moreover, for a fine quark of degree p, there exist sequences $\boldsymbol{c}, \boldsymbol{d}$, such that it holds

$$\varphi_{p,j,k} = \sum_{q=0}^{p}\sum_{l\in\mathbb{Z}} c_{p,q,k,l}\varphi_{q,j-1,l} + \sum_{q=0}^{p}\sum_{n\in\mathbb{Z}} d_{p,q,k,n}\psi_{q,j-1,n}. \qquad (4.2.13)$$

Proof. Combining Proposition 4.14, Theorem 3.2 and (3.14) we deduce the decomposition relation (4.2.12) for fine multiquarks. Moreover, since $T(z)$ is a lower triangular matrix, its inverse is of lower triangular shape as well. Hence (4.2.12) can be rewritten for single quarks as (4.2.13). $\qquad\square$

In the following we consider the length of the reconstruction sequences.

Theorem 4.16. *Let $\rho \in \{0,1\}$. The matrix*

$$\begin{pmatrix} \mathscr{C}_0(z^2) & \mathscr{D}_0(z^2) \\ \mathscr{C}_1(z^2) & \mathscr{D}_1(z^2) \end{pmatrix}$$

in (3.10) consists of Laurent polynomials only. Furthermore, the length of the recon- struction sequences $(C_{\rho+2k})_{k\in\mathbb{Z}}$, $(D_{\rho+2k})_{k\in\mathbb{Z}}$ scales linearly with p.

Proof. By (3.14), invertibility of $X(z)$ implies invertibility of the matrix

$$S(z^2) := \begin{pmatrix} \mathscr{A}_0(z^2) & \mathscr{A}_1(z^2) \\ \mathscr{B}_0(z^2) & \mathscr{B}_1(z^2) \end{pmatrix}$$

from (3.10). Furthermore, (4.2.9) implies that $\det S(z^2)$ is a monomial. Hence, since any subdeterminant of $S(z^2)$ is also a Laurent polynomial, each entry of $S(z^2)^{-1}$ is a Laurent polynomial as well. The claim follows with another application of (3.10). The dimension of the matrix $S(z^2)$ is of order p, where the length of the symbols does not depend on p, see (4.1.3), (4.2.1) . Hence the degree of its subdeterminants is linearly increasing with respect to p and the second claim follows. $\qquad\square$

Remark 4.17. So far, we have only proved the existence and estimated the length of the reconstruction sequences. Since the symbol matrix of the quarks is of lower triangular shape, we are able to inductively compute their inverse. We provide an algorithm that allows for efficient calculation of the reconstruction sequences. Suppose that for $p - 1$ the matrix $T_1(z)$ with dimension $p \times p$ has been computed. Then, for p the matrix $T(z)$ from (4.2.11) and their inverse are given by

$$T(z) = \begin{pmatrix} T_1(z) & 0 \\ x & (T(z))_{p,p} \end{pmatrix}, \quad T(z)^{-1} = \begin{pmatrix} T_1(z)^{-1} & 0 \\ y & (T(z))_{p,p}^{-1} \end{pmatrix}.$$

It is easy to verify that the last row of $T(z)^{-1}$ can be computed by

$$(T(z)^{-1})_{p,k} = y_k = -(T(z))_{p,p}^{-1} \sum_{i=k}^{p-1} x_i (T_1(z)^{-1})_{i,k}$$

$$= -\frac{1}{2^{-p}4z} \sum_{i=k}^{p-1} (T(z))_{p,i} (T_1(z)^{-1})_{i,k}, \quad k = 0,\dots,p-1.$$

Tables 4.1-4.6 show symbol matrices of multiquarks and symbol matrices of mul- tiquarklets, respectively. Additionally, the reconstruction sequences are shown. The computations have been performed with [82]. The following theorem states the exis- tence of a decomposition of a fine quark on arbitrary level j in terms of coarse quarks and quarklets. In addition we specify the number of used quarklet frame elements. We need this result in Chapter 7.

Theorem 4.18. *Each multiquark* $\Phi(2^j \cdot - \rho)$ *on level* j *has a decomposition in terms of multiquarks* Φ *and multiquarklets* Ψ, *i.e., it holds*

$$\Phi(2^j \cdot - \rho) = \sum_{k=jpk_-}^{jpk_+} C_{j,\rho+2k} \Phi(\cdot - k) + \sum_{i=0}^{j-1} \sum_{n=ipk_-+pn_-}^{ipk_++pn_+} D_{i,\rho+2n} \Psi(2^{j-1-i} \cdot - n). \quad (4.2.14)$$

Furthermore, the length of the sequences C_j (D_i) *is of order* jp (ip). *The overall length of the reconstruction sequences is of order* $j^2 p$.

Proof. To prove (4.2.14), we iteratively apply the decomposition relation (4.2.12). In particular we have to determine the number of multiquarks and multiquarklets, respectively. With the length of the reconstruction sequences being of order p, see Theorem 4.16, we conclude that the translations corresponding to the nontrivial coefficients are contained in the interval $[pk_-, pk_+]$ and $[pn_-, pn_+]$, respectively. Without loss of generality we assume $k_-, n_- < 0 < k_+, n_+$. We have

$$\Phi(2^j \cdot - \rho) = \sum_{k=pk_-}^{pk_+} C_{0,\rho+2k} \Phi(2^{j-1} \cdot - k) + \sum_{n=pn_-}^{pn_+} D_{0,\rho+2n} \Psi(2^{j-1} \cdot - n).$$

Again decomposing the multiquarks leads to

$$\sum_{k=pk_-}^{pk_+} C_{0,\rho+2k} \Phi(2^{j-1} \cdot - k) = \sum_{k=2pk_-}^{2pk_+} C_{1,\rho+2k} \Phi(2^{j-2} \cdot - k)$$

$$+ \sum_{n=pk_-+pn_-}^{pk_++pn_+} D_{1,\rho+2n} \Psi(2^{j-2} \cdot - n).$$

Inductively we get

$$\Phi(2^j \cdot - \rho) = \sum_{k=jpk_-}^{jpk_+} C_{j,\rho+2k} \Phi(\cdot - k) + \sum_{i=0}^{j-1} \sum_{n=ipk_-+pn_-}^{ipk_++pn_+} D_{i,\rho+2n} \Psi(2^{j-1-i} \cdot - n).$$

Counting the nontrivial entries of C_j, D_i leads to $|\operatorname{supp} C_j| = jp(k_+ - k_-)$, $|\operatorname{supp} D_i| = ip(k_+ - k_-) + p(n_+ - n_-)$. Since k_-, k_+, n_-, n_+ depend on the reconstruction properties of the underlying wavelet basis, only, we get $|\operatorname{supp} C_j| \sim jp$, $|\operatorname{supp} D_i| \sim ip$. Summation over i yields a total number of function vectors of order $j^2 p$. $\qquad \square$

Table 4.1: Refinement coefficients of Φ in z-notation.

m	p	$\mathscr{A}(z)$
1	0	$\begin{pmatrix} z+1 \end{pmatrix}$
	1	$\begin{pmatrix} z+1 & 0 \\ \frac{1}{2}z & \frac{1}{2}z+\frac{1}{2} \end{pmatrix}$
	2	$\begin{pmatrix} z+1 & 0 & 0 \\ \frac{1}{2}z & \frac{1}{2}z+\frac{1}{2} & 0 \\ \frac{1}{4}z & \frac{1}{2}z & \frac{1}{4}z+\frac{1}{4} \end{pmatrix}$
2	0	$\begin{pmatrix} \frac{\frac{1}{2}z^2+z+\frac{1}{2}}{z} \end{pmatrix}$
	1	$\begin{pmatrix} \frac{\frac{1}{2}z^2+z+\frac{1}{2}}{z} & 0 \\ \frac{\frac{1}{4}z^2-\frac{1}{4}}{z} & \frac{\frac{1}{4}z^2+\frac{1}{2}z+\frac{1}{4}}{z} \end{pmatrix}$
	2	$\begin{pmatrix} \frac{\frac{1}{2}z^2+z+\frac{1}{2}}{z} & 0 & 0 \\ \frac{\frac{1}{4}z^2-\frac{1}{4}}{z} & \frac{\frac{1}{4}z^2+\frac{1}{2}z+\frac{1}{4}}{z} & 0 \\ \frac{\frac{1}{8}z^2+\frac{1}{8}}{z} & \frac{\frac{1}{4}z^2-\frac{1}{4}}{z} & \frac{\frac{1}{8}z^2+\frac{1}{4}z+\frac{1}{8}}{z} \end{pmatrix}$
3	0	$\begin{pmatrix} \frac{\frac{1}{4}z^3+\frac{3}{4}z^2+\frac{3}{4}z+\frac{1}{4}}{z} \end{pmatrix}$
	1	$\begin{pmatrix} \frac{\frac{1}{4}z^3+\frac{3}{4}z^2+\frac{3}{4}z+\frac{1}{4}}{z} & 0 \\ \frac{\frac{1}{4}z^3+\frac{3}{8}z^2-\frac{1}{8}}{z} & \frac{\frac{1}{8}z^3+\frac{3}{8}z^2+\frac{3}{8}z+\frac{1}{8}}{z} \end{pmatrix}$

Table 4.2: Quarklet mask of Ψ in z-notation.

m	p	$\mathscr{B}(z)$
1	0	$\begin{pmatrix} -z+1 \end{pmatrix}$
	1	$\begin{pmatrix} -z+1 & 0 \\ 0 & -z+1 \end{pmatrix}$
	2	$\begin{pmatrix} -z+1 & 0 & 0 \\ 0 & -z+1 & 0 \\ 0 & 0 & -z+1 \end{pmatrix}$
2	0	$\begin{pmatrix} \frac{\frac{1}{4}z^4+\frac{1}{2}z^3-\frac{3}{2}z^2+\frac{1}{2}z+\frac{1}{4}}{z} \end{pmatrix}$
	1	$\begin{pmatrix} \frac{\frac{1}{4}z^4+\frac{1}{2}z^3-\frac{3}{2}z^2+\frac{1}{2}z+\frac{1}{4}}{z} & 0 \\ 0 & \frac{\frac{1}{4}z^4+\frac{1}{2}z^3-\frac{3}{2}z^2+\frac{1}{2}z+\frac{1}{4}}{z} \end{pmatrix}$
	2	$\begin{pmatrix} \frac{\frac{1}{4}z^4+\frac{1}{2}z^3-\frac{3}{2}z^2+\frac{1}{2}z+\frac{1}{4}}{z} & 0 & 0 \\ 0 & \frac{\frac{1}{4}z^4+\frac{1}{2}z^3-\frac{3}{2}z^2+\frac{1}{2}z+\frac{1}{4}}{z} & 0 \\ 0 & 0 & \frac{\frac{1}{4}z^4+\frac{1}{2}z^3-\frac{3}{2}z^2+\frac{1}{2}z+\frac{1}{4}}{z} \end{pmatrix}$
3	0	$\begin{pmatrix} \frac{\frac{3}{32}z^7+\frac{9}{32}z^6-\frac{7}{32}z^5-\frac{45}{32}z^4+\frac{45}{32}z^3+\frac{7}{32}z^2-\frac{9}{32}z-\frac{3}{32}}{z^3} \end{pmatrix}$
	1	$\begin{pmatrix} \frac{\frac{3}{32}z^7+\frac{9}{32}z^6-\frac{7}{32}z^5-\frac{45}{32}z^4+\frac{45}{32}z^3+\frac{7}{32}z^2-\frac{9}{32}z-\frac{3}{32}}{z^3} & 0 \\ 0 & \frac{\frac{3}{32}z^7+\frac{9}{32}z^6-\frac{7}{32}z^5-\frac{45}{32}z^4+\frac{45}{32}z^3+\frac{7}{32}z^2-\frac{9}{32}z-\frac{3}{32}}{z^3} \end{pmatrix}$

Table 4.3: Reconstruction coefficients of $\Phi(2\cdot)$ in z-notation.

m	p	$\mathscr{C}_0(z^2)$
1	0	$\begin{pmatrix} \frac{1}{2} \end{pmatrix}$
	1	$\begin{pmatrix} \frac{1}{2} & 0 \\ -\frac{1}{4} & 1 \end{pmatrix}$
	2	$\begin{pmatrix} \frac{1}{2} & 0 & 0 \\ -\frac{1}{4} & 1 & 0 \\ 0 & -1 & 2 \end{pmatrix}$
2	0	$\begin{pmatrix} \frac{-\frac{1}{8}z^4+\frac{3}{4}z^2-\frac{1}{8}}{z^2} \end{pmatrix}$
	1	$\begin{pmatrix} \frac{-\frac{1}{8}z^4+\frac{3}{4}z^2-\frac{1}{8}}{z^2} & 0 \\ \frac{\frac{1}{64}z^8-\frac{3}{32}z^6+\frac{3}{32}z^2-\frac{1}{64}}{z^4} & \frac{-\frac{1}{4}z^4+\frac{3}{2}z^2-\frac{1}{4}}{z^2} \end{pmatrix}$
	2	$\begin{pmatrix} \frac{-\frac{1}{8}z^4+\frac{3}{4}z^2-\frac{1}{8}}{z^2} & 0 & 0 \\ \frac{\frac{1}{64}z^8-\frac{3}{32}z^6+\frac{3}{32}z^2-\frac{1}{64}}{z^4} & \frac{-\frac{1}{4}z^4+\frac{3}{2}z^2-\frac{1}{4}}{z^2} & 0 \\ \frac{-\frac{1}{256}z^{12}+\frac{5}{128}z^{10}-\frac{15}{256}z^8-\frac{13}{64}z^6-\frac{15}{256}z^4+\frac{5}{128}z^2-\frac{1}{256}}{z^6} & \frac{\frac{1}{16}z^8-\frac{3}{8}z^6+\frac{3}{8}z^2-\frac{1}{16}}{z^4} & \frac{-\frac{1}{2}z^4+3z^2-\frac{1}{2}}{z^2} \end{pmatrix}$
3	0	$\begin{pmatrix} \frac{-\frac{9}{64}z^6+\frac{45}{64}z^4-\frac{7}{64}z^2+\frac{3}{64}}{z^4} \end{pmatrix}$
	1	$\begin{pmatrix} \frac{-\frac{9}{64}z^6+\frac{45}{64}z^4-\frac{7}{64}z^2+\frac{3}{64}}{z^4} & 0 \\ \frac{-\frac{81}{16384}z^{14}+\frac{999}{16384}z^{12}-\frac{1881}{16384}z^{10}-\frac{5865}{16384}z^8+\frac{3749}{16384}z^6-\frac{1251}{16384}z^4+\frac{261}{16384}z^2-\frac{27}{16384}}{z^8} & \frac{-\frac{9}{32}z^6+\frac{45}{32}z^4-\frac{7}{32}z^2+\frac{3}{32}}{z^4} \end{pmatrix}$

Table 4.4: Reconstruction coefficients of $\Phi(2\cdot)$ in z-notation.

m	p	$\mathscr{D}_0(z^2)$
1	0	$\begin{pmatrix} \frac{1}{2} \end{pmatrix}$
	1	$\begin{pmatrix} \frac{1}{2} & 0 \\ \frac{1}{4} & \frac{1}{2} \end{pmatrix}$
	2	$\begin{pmatrix} \frac{1}{2} & 0 & 0 \\ \frac{1}{4} & \frac{1}{2} & 0 \\ 0 & \frac{1}{2} & \frac{1}{2} \end{pmatrix}$
2	0	$\begin{pmatrix} \frac{\frac{1}{4}z^2+\frac{1}{4}}{z^2} \end{pmatrix}$
	1	$\begin{pmatrix} \frac{\frac{1}{4}z^2+\frac{1}{4}}{z^2} & 0 \\ \frac{-\frac{1}{32}z^6+\frac{7}{32}z^4-\frac{7}{32}z^2+\frac{1}{32}}{z^4} & \frac{\frac{1}{4}z^2+\frac{1}{4}}{z^2} \end{pmatrix}$
	2	$\begin{pmatrix} \frac{\frac{1}{4}z^2+\frac{1}{4}}{z^2} & 0 & 0 \\ \frac{-\frac{1}{32}z^6+\frac{7}{32}z^4-\frac{7}{32}z^2+\frac{1}{32}}{z^4} & \frac{\frac{1}{4}z^2+\frac{1}{4}}{z^2} & 0 \\ \frac{\frac{1}{128}z^{10}-\frac{11}{128}z^8+\frac{13}{64}z^6+\frac{13}{64}z^4-\frac{11}{128}z^2+\frac{1}{128}}{z^6} & \frac{-\frac{1}{16}z^6+\frac{7}{16}z^4-\frac{7}{16}z^2+\frac{1}{16}}{z^4} & \frac{\frac{1}{4}z^2+\frac{1}{4}}{z^2} \end{pmatrix}$
3	0	$\begin{pmatrix} \frac{\frac{3}{8}z^2+\frac{1}{8}}{z^2} \end{pmatrix}$
	1	$\begin{pmatrix} \frac{\frac{3}{8}z^2+\frac{1}{8}}{z^2} & 0 \\ \frac{\frac{27}{2048}z^{10}-\frac{189}{2048}z^8+\frac{159}{1024}z^6-\frac{93}{1024}z^4+\frac{39}{2048}z^2-\frac{9}{2048}}{z^6} & \frac{\frac{3}{8}z^2+\frac{1}{8}}{z^2} \end{pmatrix}$

Table 4.5: Reconstruction coefficients of $\Phi(2\cdot-1)$ in z-notation.

m	p	$\mathscr{C}_1(z^2)$
1	0	$\left(\frac{1}{2}\right)$
	1	$\begin{pmatrix} \frac{1}{2} & 0 \\ -\frac{1}{4} & 1 \end{pmatrix}$
	2	$\begin{pmatrix} \frac{1}{2} & 0 & 0 \\ -\frac{1}{4} & 1 & 0 \\ 0 & -1 & 2 \end{pmatrix}$
2	0	$\left(\frac{1}{4}z^2+\frac{1}{4}\right)$
	1	$\begin{pmatrix} \frac{1}{4}z^2+\frac{1}{4} & 0 \\ \frac{-\frac{1}{32}z^6-\frac{1}{32}z^4+\frac{1}{32}z^2+\frac{1}{32}}{z^2} & \frac{1}{2}z^2+\frac{1}{2} \end{pmatrix}$
	2	$\begin{pmatrix} \frac{1}{4}z^2+\frac{1}{4} & 0 & 0 \\ \frac{-\frac{1}{32}z^6-\frac{1}{32}z^4+\frac{1}{32}z^2+\frac{1}{32}}{z^2} & \frac{1}{2}z^2+\frac{1}{2} & 0 \\ \frac{\frac{1}{128}z^{10}-\frac{3}{128}z^8-\frac{7}{64}z^6-\frac{7}{64}z^4-\frac{3}{128}z^2+\frac{1}{128}}{z^4} & \frac{-\frac{1}{8}z^6-\frac{1}{8}z^4+\frac{1}{8}z^2+\frac{1}{8}}{z^2} & z^2+1 \end{pmatrix}$
3	0	$\left(\frac{\frac{3}{64}z^6-\frac{7}{64}z^4+\frac{45}{64}z^2-\frac{9}{64}}{z^2}\right)$
	1	$\begin{pmatrix} \frac{\frac{3}{64}z^6-\frac{7}{64}z^4+\frac{45}{64}z^2-\frac{9}{64}}{z^2} & 0 \\ \frac{\frac{27}{16384}z^{14}-\frac{261}{16384}z^{12}+\frac{483}{16384}z^{10}-\frac{1957}{16384}z^8-\frac{5655}{16384}z^6+\frac{4185}{16384}z^4-\frac{999}{16384}z^2+\frac{81}{16384}}{z^6} & \frac{\frac{3}{32}z^6-\frac{7}{32}z^4+\frac{45}{32}z^2-\frac{9}{32}}{z^2} \end{pmatrix}$

Table 4.6: Reconstruction coefficients of $\Phi(2\cdot-1)$ in z-notation.

m	p	$\mathscr{D}_1(z^2)$
1	0	$\left(-\frac{1}{2}\right)$
	1	$\begin{pmatrix} -\frac{1}{2} & 0 \\ \frac{1}{4} & -\frac{1}{2} \end{pmatrix}$
	2	$\begin{pmatrix} -\frac{1}{2} & 0 & 0 \\ \frac{1}{4} & -\frac{1}{2} & 0 \\ 0 & \frac{1}{2} & -\frac{1}{2} \end{pmatrix}$
2	0	$\left(-\frac{1}{2}\right)$
	1	$\begin{pmatrix} -\frac{1}{2} & 0 \\ \frac{\frac{1}{16}z^4-\frac{1}{16}}{z^2} & -\frac{1}{2} \end{pmatrix}$
	2	$\begin{pmatrix} -\frac{1}{2} & 0 & 0 \\ \frac{\frac{1}{16}z^4-\frac{1}{16}}{z^2} & -\frac{1}{2} & 0 \\ \frac{-\frac{1}{64}z^8+\frac{1}{16}z^6+\frac{5}{32}z^4+\frac{1}{16}z^2-\frac{1}{64}}{z^4} & \frac{\frac{1}{8}z^4-\frac{1}{8}}{z^2} & -\frac{1}{2} \end{pmatrix}$
3	0	$\left(-\frac{1}{8}z^2-\frac{3}{8}\right)$
	1	$\begin{pmatrix} -\frac{1}{8}z^2-\frac{3}{8} & 0 \\ \frac{-\frac{9}{2048}z^{10}+\frac{39}{2048}z^8-\frac{93}{1024}z^6+\frac{159}{1024}z^4-\frac{189}{2048}z^2+\frac{27}{2048}}{z^4} & -\frac{1}{8}z^2-\frac{3}{8} \end{pmatrix}$

4.3 Quarklets on the Interval

In this section we discuss the construction of quarklets on the interval. For the inner quarklets, we proceed as before and assign the two-scale relation of the underlying wavelet to the quarks with the same coefficients. Quite surprisingly, a similar approach does not work for the boundary quarklets since this would destroy the vanishing moment properties. It turns out that in order to preserve the vanishing moment properties of the underlying wavelet Riesz basis for the full quarklet system, it is necessary to define the two-scale relation of the boundary quarklets appropriately. In any case, quarklets are defined as linear combinations of quark generators on the next higher level. Then, the relation (4.2.1) for one quarklet becomes

$$\psi_{p,j,k}^{\vec{\sigma}} := \sum_{l \in \Delta_{j+1,\vec{\sigma}}} b_{k,l}^{p,j,\vec{\sigma}} \varphi_{p,j+1,l}, \quad k \in \nabla_{j,\vec{\sigma}}, \tag{4.3.1}$$

with $\Delta_{j,\vec{\sigma}} \subset \mathbb{Z}$ as in (2.2.29) and $\nabla_{j,\vec{\sigma}}$ defined by

$$\nabla_{j,\vec{\sigma}} := \begin{cases} \Delta_{j,\vec{\sigma}}, & j = j_0 - 1, \\ \Delta_{j+1,\vec{\sigma}} \setminus \Delta_{j,\vec{\sigma}}, & j \geq j_0, \end{cases} \tag{4.3.2}$$

for appropriately chosen j_0. We already notice that in contrast to (4.2.1) the coefficients $b_{k,l}^{p,j,\vec{\sigma}}$ in (4.3.1) do not only depend on the translation parameter l. At first, let us discuss the construction of the inner quarklets. For $p, j \in \mathbb{N}_0$, $j \geq j_0$, $k \in \nabla_{j,\vec{\sigma}}$ with $m - 1 \leq k \leq 2^j - m$ the inner wavelets of the Primbs basis are given by $\psi_{j,k}^{\vec{\sigma}} = \sum_{l \in \Delta_{j+1,\vec{\sigma}}} b_{k,l}^{j,\vec{\sigma}} \varphi_{j+1,l}$, cf. (2.2.14). We construct an inner quarklet by keeping these coefficients and inserting them into (4.3.1):

$$b_{k,l}^{p,j,\vec{\sigma}} := b_{k,l}^{j,\vec{\sigma}}, \quad m - 1 \leq k \leq 2^j - m, l \in \Delta_{j+1,\vec{\sigma}}. \tag{4.3.3}$$

If the inner Primbs wavelets have $\widetilde{m}$ vanishing moments, the inner quarklets defined above have the same number of vanishing moments. This result is shown in [28] for cardinal B-spline quarks and therefore it holds true for the inner Schoenberg B-spline quarks.

Lemma 4.19. *[28, Lem. 2] For each $p \in \mathbb{N}_0$, the quarklet ψ_p has $\widetilde{m}$ vanishing moments.*

The next step is to construct boundary quarklets. As already mentioned, the coefficients of the boundary wavelets are not suitable for the boundary quarklets, since in general the vanishing moment properties can not be preserved. A simple counter-example is given by

$$\int_{\mathbb{R}} \sum_{l=-1}^{2} b_{0,l}^{2,(0,0)} \varphi_{1,3,l} = \frac{1}{8},$$

where the non-trivial coefficients are $(b_{0,l}^{2,(0,0)})_{l=-1}^2 = \sqrt{2}(\frac{3}{2}, -\frac{9}{8}, \frac{1}{4}, \frac{1}{8})$. Therefore, Instead of keeping the coefficients, our approach is to modify the coefficients in that way that the $\widetilde{m}$ equations

$$\int_{\mathbb{R}} x^q \psi_{p,j,k}^{\vec{\sigma}}(x) \, dx = \int_{\mathbb{R}} x^q \sum_{l \in \Delta_{j+1,\vec{\sigma}}} b_{k,l}^{p,j,\vec{\sigma}} \varphi_{p,j+1,l}(x) \, dx = 0, \quad q = 0, ..., \widetilde{m} - 1. \quad (4.3.4)$$

are fulfilled not only for $p = 0$ but for all $p \in \mathbb{N}_0$. We restrict our discussion to left boundary quarklets, i.e., $k = 0, \ldots, m-2$, and assume that they are only composed of left boundary and inner quarks. To get at least one non-trivial solution of (4.3.4) we further assume that every boundary quarklet consists of $\widetilde{m} + 1$ quarks. Furthermore the k-th quarklets representation should begin at the leftmost but k-th quark with respect to boundary conditions. This leads to the $\widetilde{m} \times (\widetilde{m} + 1)$ linear system of equations

$$\sum_{l=-m+1+\operatorname{sgn}\sigma_l+k}^{-m+1+\operatorname{sgn}\sigma_l+k+\widetilde{m}} b_{k,l}^{p,j,\vec{\sigma}} \int_{\mathbb{R}} x^q \varphi_{p,j+1,l}(x) \, dx = 0, \quad q = 0, ..., \widetilde{m} - 1. \quad (4.3.5)$$

Observe that the associated coefficient matrix is a rectangular matrix of size $\widetilde{m} \times (\widetilde{m} + 1)$ and has a non-trivial kernel, so that we can find solutions to (4.3.5). We always choose one solution with ℓ_2 norm equal to 1. Hence, we are able to construct quarklets at the boundary with vanishing moments.

Definition 4.20. If $0 \neq b_k^{p,j,\vec{\sigma}} \in \mathbb{R}^{\widetilde{m}+1}$ solves (4.3.5), we define the k-th left boundary quarklet by

$$\psi_{p,j,k}^{\vec{\sigma}} := \sum_{l=-m+1+\operatorname{sgn}\sigma_l+k}^{-m+1+\operatorname{sgn}\sigma_l+k+\widetilde{m}} b_{k,l}^{p,j,\vec{\sigma}} \varphi_{p,j+1,l}, \quad k = 0, \ldots, m-2. \quad (4.3.6)$$

Figures 4.6-4.10 show inner and left boundary quarklets. The vanishing moment property of the quarklets immediately leads to the following cancellation property of the quarklets.

Lemma 4.21. *Let $p, j \in \mathbb{N}_0$, $j \geq j_0$, $k \in \nabla_{j,\vec{\sigma}}$ and $\psi_{p,j,k}^{\vec{\sigma}}$ be a quarklet with $\widetilde{m}$ vanishing moments. There exists a constant $C(m, \psi) > 0$, such that for every $r \in \mathbb{N}_0, r \leq \widetilde{m} - 1$ and $f \in W_\infty^r(\mathbb{R})$:*

$$|\langle f, \psi_{p,j,k}^{\vec{\sigma}} \rangle_{L_2(\mathbb{R})}| \leq C(p+1)^{-m} 2^{-j(r+1/2)} |f|_{W_\infty^r(\operatorname{supp} \psi_{p,j,k}^{\vec{\sigma}})}. \quad (4.3.7)$$

Proof. The proof can be performed by following the lines of [28, Lem. 3]. From the vanishing moments of the quarklets, Hölder's inequality and a Whitney type estimate it follows:

$$|\langle f, \psi_{p,j,k}^{\vec{\sigma}} \rangle_{L_2(\mathbb{R})}| \leq C_1 |\operatorname{supp} \psi_{p,j,k}|^r |f|_{W_\infty^r(\operatorname{supp} \psi_{p,j,k}^{\vec{\sigma}})} \|\psi_{p,j,k}^{\vec{\sigma}}\|_{L_1(\operatorname{supp} \psi_{p,j,k}^{\vec{\sigma}})}, \quad (4.3.8)$$

where $C_1 > 0$ only depends on r. To further estimate the L_1-norm expression in (4.3.8) we use the symmetry of the boundary quarks, (4.3.1) and the relation

$$\varphi_{p,j,k} = 2^{j/2}\varphi_{p,0,k}(2^j\cdot), \quad k = -m+1, \ldots, 2^j - m.$$

Combining this relation and the norm estimates (4.1.18), (4.1.13) we obtain

$$\|\psi_{p,j,k}^{\vec{\sigma}}\|_{L_1(\operatorname{supp}\psi_{p,j,k}^{\vec{\sigma}})} \leq C_2 2^{-\frac{j+1}{2}}(p+1)^{-m} \sum_{l\in\Delta_{j+1,\vec{\sigma}}} |b_{k,l}^{p,j,\vec{\sigma}}|,$$

where $C_2 > 0$ only depends on m. The claim finally follows by estimating the asymptotic behaviour of $|\operatorname{supp}\psi_{p,j,k}^{\vec{\sigma}}|$ by 2^{-j}. $\qquad\square$

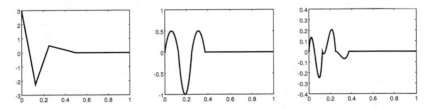

Figure 4.6: Left boundary quarklet for $m = 2$ and $p = 0, 1, 2$.

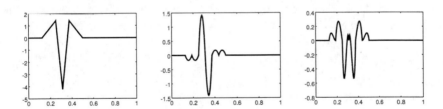

Figure 4.7: Inner quarklet for $m = 2$ and $p = 0, 1, 2$.

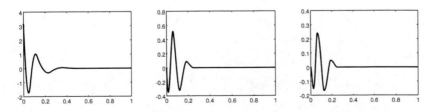

Figure 4.8: First left boundary quarklet for $m = 3$ and $p = 0, 1, 2$.

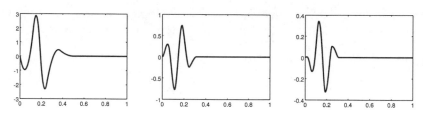

Figure 4.9: Second left boundary quarklet for $m = 3$ and $p = 0, 1, 2$.

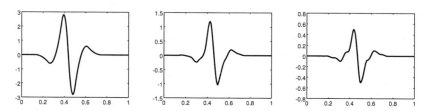

Figure 4.10: Inner quarklet for $m = 3$ and $p = 0, 1, 2$.

4.4 Quarklet Frames for L_2

In this section we state the frame property for quarklet systems in the space $L_2(I)$, see Theorem 4.23. We restrict ourselves to the interval case, nonetheless the result holds true for the shift-invariant case, cf. [28,57]. We will need the L_2 frame property in Chapter 5. The following proposition transfers the estimates for the Gramian matrices from [28, Prop. 2] to the boundary adapted case. This is the last missing ingredient to show the frame property of the quarklet systems in $L_2(I)$.

Proposition 4.22. *For fixed $p \in \mathbb{N}_0$, the operators induced by the Gramian matrices, which are given by*

$$G_p := \left(\langle \varphi_{p,j_0,k}, \varphi_{p,j_0,k'} \rangle_{L_2(\mathbb{R})} \right)_{k,k' \in \nabla_{j_0-1,\vec{\sigma}}}, \tag{4.4.1}$$

$$H_p := \left(\langle \psi_{p,j,k}^{\vec{\sigma}}, \psi_{p,j',k'}^{\vec{\sigma}} \rangle_{L_2(\mathbb{R})} \right)_{(j,k):j \geq j_0, k \in \nabla_{j,\vec{\sigma}}, (j',k'):j' \geq j_0, k' \in \nabla_{j',\vec{\sigma}}} \tag{4.4.2}$$

are bounded operators on $\ell_2(\{(j_0 - 1, k) : k \in \nabla_{j_0-1,\vec{\sigma}}\})$ and $\ell_2(\{(j,k) : j \geq j_0, k \in \nabla_{j,\vec{\sigma}}\}))$, respectively, i.e., there exist constants $C' = C'(m, \varphi) > 0$, $C'' = C''(m, \psi) > 0$, such that

$$\|G_p\|_{\mathcal{L}(\ell_2(\{(j_0-1,k):k \in \nabla_{j_0-1,\vec{\sigma}}\}))} \leq C'(p+1)^{-(2m-1)}, \tag{4.4.3}$$

$$\|H_p\|_{\mathcal{L}(\ell_2(\{(j,k):j \geq j_0, k \in \nabla_{j,\vec{\sigma}}\}))} \leq C''(p+1)^{-1}. \tag{4.4.4}$$

Proof. The proof is based upon the cancellation property (4.3.7) and can be performed by following the lines of [28, Prop. 2]. □

After introducing the construction of quarks and quarklets on the interval and proving some crucial estimates we are finally able to transfer the frame properties of the shift-invariant quarklets, to the case of boundary adapted quarklets. The notation

$$\psi_{p,j_0-1,k}^{\vec{\sigma}} := \varphi_{p,j_0,k}, \quad p \in \mathbb{N}_0, k \in \nabla_{j_0-1,\vec{\sigma}}. \tag{4.4.5}$$

allows for a convenient description of quarkonial systems. We formulate the frame property in $L_2(I)$.

Theorem 4.23. *Let the boundary adapted quarks and quarklets be defined by (4.1.2), (4.3.1). Let the index sets $\nabla_{j,\vec{\sigma}}$ be defined by (4.3.2). We define the index set for whole quarklet systems on the interval by*

$$\nabla_{\vec{\sigma}} := \{(p, j, k) : p, j \in \mathbb{N}_0, j \geq j_0 - 1, k \in \nabla_{j,\vec{\sigma}}\}. \tag{4.4.6}$$

The weighted quarklet system

$$\Psi_{\vec{\sigma}} := \{(p+1)^{-\delta/2} \psi_{p,j,k}^{\vec{\sigma}} : (p, j, k) \in \nabla_{\vec{\sigma}}\}, \quad \delta > 1, \tag{4.4.7}$$

is a frame for $L_2(I)$.

Proof. The used weights $w_p := (p+1)^{-\delta/2}$ fulfil $w_0 = 1$, hence the quarklet system contains an underlying Riesz basis which implies the lower frame estimate, cf. [28, Thm. 3]. The convergence of the sum $\sum_p w_p(p+1)^{-1/2} < \infty$ implies the upper frame estimate, see again [28, Thm. 3]. □

4.5 Characterisation of Besov Spaces for $q \geq 1$

In this section we discuss stability properties of quarklet systems in scales of Besov spaces. This can be interpreted as a generalisation of the frame property in Sobolev spaces discussed in [26, 28, 57]. We restrict our discussion to the shift-invariant case, nonetheless the claims made in this section can be formulated for the interval case with the usual change of notation. It is well-known that a wavelet basis of $L_2(\mathbb{R})$ with certain properties characterises Besov spaces in the following sense: For functions f equivalent norms are given by

$$\|f\|_{B_q^s(L_q(\mathbb{R}))}^q \sim \sum_{j \geq -1} \sum_{k \in \mathbb{Z}} 2^{j(s+\frac{1}{2}-\frac{1}{q})q} |\langle f, \psi_{j,k} \rangle|^q, \tag{4.5.1}$$

$$\|f\|_{B_r^s(L_q(\mathbb{R}))}^r \sim \sum_{j \geq -1} 2^{j(s+\frac{1}{2}-\frac{1}{q})r} \left(\sum_{k \in \mathbb{Z}} |\langle f, \psi_{j,k} \rangle|^q \right)^{r/q}. \tag{4.5.2}$$

In the sequel we derive characterisations of Besov spaces in terms of quarklet systems. Since these systems contain the underlying wavelet basis, it suffices to prove the corresponding upper estimates. The proofs presented in this section rely on ideas from [30] and on classical wavelet techniques, cf. [92, Chap. 8-9]. We start with a first stability result for quark generators in L_q spaces.

Proposition 4.24. *Let $\hat{p} \in \mathbb{N}_0$, $1 \leq q < \infty$. Let the cardinal B-spline quark φ_p be defined by (4.1.1). Then the following upper estimate holds true for all sequences $c = \{c_{p,k}\}_{p=0,\ldots,\hat{p},k\in\mathbb{Z}}$:*

$$\|\sum_{p=0}^{\hat{p}}\sum_{k\in\mathbb{Z}} c_{p,k}\varphi_p(\cdot - k)\|_{L_q(\mathbb{R})} \lesssim (\hat{p}+1)^{1-\frac{1}{q}} \left(\sum_{p=0}^{\hat{p}}\sum_{k\in\mathbb{Z}} |c_{p,k}|^q\right)^{1/q}. \tag{4.5.3}$$

Proof. Using the triangle inequality and with $\frac{1}{q} + \frac{1}{t} = 1$ we get

$$\|\sum_{\substack{p=0,\ldots,\hat{p} \\ k\in\mathbb{Z}}} c_{p,k}\varphi_p(\cdot - k)\|_{L_q(\mathbb{R})}^q = \int_{\mathbb{R}} |\sum_{p=0,\ldots,\hat{p},\, k\in\mathbb{Z}} c_{p,k}\varphi_p(x-k)|^q \mathrm{d}x$$

$$\leq \int_{\mathbb{R}} \left(\sum_{p=0,\ldots,\hat{p},\, k\in\mathbb{Z}} |c_{p,k}||\varphi_p(x-k)|^{\frac{1}{q}}|\varphi_p(x-k)|^{\frac{1}{t}}\right)^q \mathrm{d}x.$$

Using Hölder's inequality for series yields

$$\|\sum_{\substack{p=0,\ldots,\hat{p} \\ k\in\mathbb{Z}}} c_{p,k}\varphi_p(\cdot - k)\|_{L_q(\mathbb{R})}^q \leq \int_{\mathbb{R}} \left(\sum_{p=0,\ldots,\hat{p},\, k\in\mathbb{Z}} |c_{p,k}|^q|\varphi_p(x-k)|\right)$$

$$\cdot \left(\sum_{p'=0,\ldots,\hat{p},\, k'\in\mathbb{Z}} |\varphi_{p'}(x-k')|\right)^{q-1} \mathrm{d}x.$$

Due to the compact support of φ_p we have

$$\|\sum_{\substack{p=0,\ldots,\hat{p} \\ k\in\mathbb{Z}}} c_{p,k}\varphi_p(\cdot - k)\|_{L_q(\mathbb{R})}^q \leq \sum_{\substack{p=0,\ldots,\hat{p} \\ k\in\mathbb{Z}}} |c_{p,k}|^q \int_k^{k+m} |\varphi_p(x-k)|$$

$$\cdot \left(\sum_{p'=0,\ldots,\hat{p},\, k'\in\mathbb{Z}} |\varphi_{p'}(x-k')|\right)^{q-1} \mathrm{d}x.$$

Now with Hölder's inequality for integrals and the norm estimate (4.1.18) we obtain

$$\|\sum_{\substack{p=0,\dots,\widehat{p} \\ k\in\mathbb{Z}}} c_{p,k}\varphi_p(\cdot - k)\|^q_{L_q(\mathbb{R})} \leq \sum_{\substack{p=0,\dots,\widehat{p},\, k\in\mathbb{Z}}} |c_{p,k}|^q \|\varphi_p(\cdot - k)\|_{L_1(\mathbb{R})}$$

$$\sum_{p'=0,\dots,\widehat{p},\, k'\in\mathbb{Z}} \|\varphi_{p'}(\cdot - k')\|^{q-1}_{L_\infty(k,k+m)}$$

$$\leq \sum_{\substack{p=0,\dots,\widehat{p} \\ k\in\mathbb{Z}}} |c_{p,k}|^q \|\sum_{\substack{p'=0,\dots,\widehat{p} \\ k'\in\mathbb{Z}}} \varphi_{p'}(\cdot - k')\|^{q-1}_{L_\infty(k,k+m)}$$

$$\leq \sum_{\substack{p=0,\dots,\widehat{p} \\ k\subset\mathbb{Z}}} |c_{p,k}|^q \left(\sum_{\substack{p'=0,\dots,\widehat{p} \\ k'\in\mathbb{Z}}} \|\varphi_{p'}(\cdot - k')\|_{L_\infty(k,k+m)} \right)^{q-1} .$$

The number of B-Splines whose supports intersect with $[k, k + m]$ is bounded by a constant C just depending on m, thus the number of quarks is bounded by $(\widehat{p} + 1)C$ and the claim follows:

$$\|\sum_{\substack{p=0,\dots,\widehat{p} \\ k\in\mathbb{Z}}} c_{p,k}\varphi_p(\cdot - k)\|^q_{L_q(\mathbb{R})} \lesssim (\widehat{p} + 1)^{q-1} \sum_{\substack{p=0,\dots,\widehat{p} \\ k\in\mathbb{Z}}} |c_{p,k}|^q.$$

$\square$

Next we show an analogous stability result for single-scale quarklet systems.

Proposition 4.25. *Let $\widehat{p} \in \mathbb{N}_0$, $1 \leq q < \infty$ and let $j \geq 0$ be fixed. Let the quarklets $\psi_{p,j,k}$ be defined by (4.2.2). Then the following upper estimate holds true for all sequences $c = \{c_{p,j,k}\}_{p=0,\dots,\widehat{p},k\in\mathbb{Z}}$:*

$$\|\sum_{p=0}^{\widehat{p}}\sum_{k\in\mathbb{Z}} c_{p,j,k}\psi_{p,j,k}\|_{L_q(\mathbb{R})} \lesssim (\widehat{p} + 1)^{1-\frac{1}{q}} 2^{j(\frac{1}{2}-\frac{1}{q})} \left(\sum_{p=0}^{\widehat{p}}\sum_{k\in\mathbb{Z}} |c_{p,j,k}|^q \right)^{1/q} . \qquad (4.5.4)$$

Proof. With the definition of the quarklets $\psi_{p,j,k}$ in (4.2.2) we obtain

$$\|\sum_{\substack{p=0,\dots,\widehat{p} \\ k\in\mathbb{Z}}} c_{p,j,k}\psi_{p,j,k}\|^q_{L_q(\mathbb{R})} = \|\sum_{\substack{p=0,\dots,\widehat{p} \\ k\in\mathbb{Z}}} c_{p,j,k}2^{j/2}\psi_p(2^j \cdot -k)\|^q_{L_q(\mathbb{R})}.$$

Substitution leads to

$$\|\sum_{\substack{p=0,\dots,\widehat{p} \\ k\in\mathbb{Z}}} c_{p,j,k}\psi_{p,j,k}\|^q_{L_q(\mathbb{R})} \leq 2^{j(\frac{1}{2}-\frac{1}{q})q} \|\sum_{\substack{p=0,\dots,\widehat{p} \\ k\in\mathbb{Z}}} c_{p,j,k}\psi_p(\cdot - k)\|^q_{L_q(\mathbb{R})}.$$

Following the lines of the proof of Proposition 4.24, we get

$$\Big\| \sum_{\substack{p=0,\ldots,\widehat{p} \\ k\in\mathbb{Z}}} c_{p,j,k}\psi_{p,j,k} \Big\|_{L_q(\mathbb{R})}^q \leq 2^{j(\frac{1}{2}-\frac{1}{q})q} \sum_{\substack{p=0,\ldots,\widehat{p} \\ k\in\mathbb{Z}}} |c_{p,j,k}|^q \|\psi_p(\cdot-k)\|_{L_1(\mathbb{R})}$$

$$\cdot \Big\| \sum_{\substack{p'=0,\ldots,\widehat{p} \\ k'\in\mathbb{Z}}} \psi_{p'}(\cdot-k') \Big\|_{L_\infty(ck,C(k+m))}^{q-1},$$

where the support of a quarklet is up to a constant determined the support of a quark generator. Combining this with the two-scale relation (4.2.1) leads to

$$\Big\| \sum_{\substack{p=0,\ldots,\widehat{p} \\ k\in\mathbb{Z}}} c_{p,j,k}\psi_{p,j,k} \Big\|_{L_q(\mathbb{R})}^q \leq 2^{j(\frac{1}{2}-\frac{1}{q})q} \sum_{\substack{p'=0,\ldots,\widehat{p} \\ k'\in\mathbb{Z}}} |c_{p,j,k}|^q \Big\| \sum_{l\in\mathbb{Z}} b_l\varphi_p(2(\cdot-k)-l) \Big\|_{L_1(\mathbb{R})}$$

$$\cdot \Big\| \sum_{\substack{p'=0,\ldots,\widehat{p} \\ k'\in\mathbb{Z}}} \sum_{l'\in\mathbb{Z}} b_l\varphi_{p'}(2(\cdot-k')-l') \Big\|_{L_\infty(ck,C(k+m))}^{q-1}.$$

Since the support of b is finite, the number of quarks with support in $[ck, C(k+m)]$ is bounded by a constant just depending on m. Finally we obtain the claim with an additional constant depending on the coefficient mask b:

$$\Big\| \sum_{\substack{p=0,\ldots,\widehat{p} \\ k\in\mathbb{Z}}} c_{p,j,k}\psi_{p,j,k} \Big\|_{L_q(\mathbb{R})}^q \lesssim 2^{j(\frac{1}{2}-\frac{1}{q})q} \sum_{\substack{p=0,\ldots,\widehat{p} \\ k\in\mathbb{Z}}} |c_{p,j,k}|^q \left(\sum_{l\in\mathbb{Z}} |b_l| \right) \|\varphi_p\|_{L_1(\mathbb{R})}$$

$$\cdot \Big\| \sum_{\substack{p'=0,\ldots,\widehat{p} \\ k'\in\mathbb{Z}}} b_{l'-k'}\varphi_{p'}(\cdot-k') \Big\|_{L_\infty(k,k+m)}^{q-1}$$

$$\lesssim 2^{j(\frac{1}{2}-\frac{1}{q})q}(\widehat{p}+1)^{q-1} \sum_{\substack{p=0,\ldots,\widehat{p} \\ k\in\mathbb{Z}}} |c_{p,j,k}|^q.$$

$$\qquad\qquad\qquad\qquad\qquad\qquad\qquad\qquad\qquad\qquad\qquad\qquad\qquad\qquad\square$$

Next we show a stability result for truncated quarklet systems.

Theorem 4.26. *Define the truncated quarklet system by*

$$\Psi_{\widehat{p}} := \{\psi_{p,j,k} : 0 \leq p \leq \widehat{p}, j \geq -1, k \in \mathbb{Z}\}. \tag{4.5.5}$$

Then, for $s < t < m-1+\frac{1}{q}$ we have the following norm estimate:

$$\|f\|_{B_q^s(L_q(\mathbb{R}))} \lesssim \inf_{(4.5.7)} (\widehat{p}+1)^{2t+1-\frac{1}{q}} \left(\sum_{p=0}^{\widehat{p}} \sum_{j\geq-1} \sum_{k\in\mathbb{Z}} 2^{j(s+\frac{1}{2}-\frac{1}{q})q} |c_{p,j,k}|^q \right)^{1/q}, \tag{4.5.6}$$

$$c: f = \sum_{p=0}^{\widehat{p}} \sum_{j\geq-1} \sum_{k\in\mathbb{Z}} c_{p,j,k}\psi_{p,j,k}. \tag{4.5.7}$$

Proof. Consider an arbitrary series

$$f = \sum_{j \geq -1} f_{\widehat{p},j} = \sum_{p=0}^{\widehat{p}} \sum_{j \geq -1} \sum_{k \in \mathbb{Z}} c_{p,j,k} \psi_{p,j,k}.$$

We use definition (1.1.29) of the Besov norm: First we consider the modulus of smoothness ω_m. Using the triangle inequality, (1.1.29) for $s < t < m - 1 + \frac{1}{q}$, $j \leq l$ and $\omega_m(f, 2^{-l}) \leq 2^m \|f\|_{L_q(\mathbb{R})}$ for $j > l$ we get

$$\omega_m(f, 2^{-l})_{L_q(\mathbb{R})} \leq \sum_{j \leq -1} \omega_m(f_{\widehat{p},j}, 2^{-l})_{L_q(\mathbb{R})}$$

$$\lesssim \sum_{j < l} 2^{-lt} \|f_{\widehat{p},j}\|_{B_q^t(L_q(\mathbb{R}))} + \sum_{j \geq l} \|f_{\widehat{p},j}\|_{L_q(\mathbb{R})}.$$

Since the functions $f_{\widehat{p},j}$ are contained in $V_{\widehat{p},j+1}$, with the Bernstein estimate (4.1.22) we obtain

$$\omega_m(f, 2^{-l})_{L_q(\mathbb{R})} \lesssim \sum_{j < l} 2^{-lt} (\widehat{p}+1)^{2t} 2^{jt} \|f_{\widehat{p},j}\|_{L_q(\mathbb{R})} + \sum_{j \geq l} \|f_{\widehat{p},j}\|_{L_q(\mathbb{R})}$$

$$= \sum_{j < l} 2^{-(l-j)t} (\widehat{p}+1)^{2t} \|f_{\widehat{p},j}\|_{L_q(\mathbb{R})} + \sum_{j \geq l} \|f_{\widehat{p},j}\|_{L_q(\mathbb{R})}.$$

Combining this estimate with (1.1.29) leads to

$$\|f\|_{B_q^s(L_q(\mathbb{R}))}^q \lesssim \sum_{l=0}^{\infty} 2^{lsq} \left(\sum_{j<l} 2^{-(l-j)t} (\widehat{p}+1)^{2t} \|f_{\widehat{p},j}\|_{L_q(\mathbb{R})} + \sum_{j \geq l} \|f_{\widehat{p},j}\|_{L_q(\mathbb{R})} \right)^q$$

$$= \sum_{l=0}^{\infty} 2^{lsq} \left(\sum_{j<l} \frac{2^{-(l-j)(t-\varepsilon)}}{2^{-(l-j)(-\varepsilon)}} (\widehat{p}+1)^{2t} \|f_{\widehat{p},j}\|_{L_q(\mathbb{R})} \right.$$

$$\left. + \sum_{j \geq l} \frac{2^{(j-l)\varepsilon}}{2^{(j-l)\varepsilon}} \|f_{\widehat{p},j}\|_{L_q(\mathbb{R})} \right)^q,$$

where $0 < \varepsilon < \min(s, t - s)$. The next step is the application of Jensen's inequality, cf. [72]. With $C_\varepsilon = \frac{2^\varepsilon - 2^{-\varepsilon(l+1)}+1}{2^\varepsilon - 1} = \sum_{j \leq l} 2^{-\varepsilon(l-j)} + \sum_{j > l} 2^{-\varepsilon(j-l)}$ the inner sum forms a convex combination with $\lambda_j = C_\varepsilon^{-1} 2^{-\varepsilon(l-j)}$, $\lambda_j = C_\varepsilon^{-1} 2^{-\varepsilon(j-l)}$, respectively. We obtain

$$\|f\|_{B_q^s(L_q(\mathbb{R}))}^q \lesssim \sum_{l=0}^{\infty} 2^{lsq} C_\varepsilon^q \left(\sum_{j<l} \lambda_j 2^{-(l-j)(t-\varepsilon)} (\widehat{p}+1)^{2t} \|f_{\widehat{p},j}\|_{L_q(\mathbb{R})} \right.$$

$$\left. + \sum_{j \geq l} \lambda_j 2^{(j-l)\varepsilon} \|f_{\widehat{p},j}\|_{L_q(\mathbb{R})} \right)^q.$$

Since $q > 1$, $(\cdot)^q$ is convex and with Jensen's inequality it follows

$$\|f\|^q_{B^s_q(L_q(\mathbb{R}))} \lesssim \sum_{l=0}^{\infty} 2^{lsq} C^q_\varepsilon \left(\sum_{j<l} \lambda_j 2^{-(l-j)(t-\varepsilon)q} (\widehat{p}+1)^{2tq} \|f_{\widehat{p},j}\|^q_{L_q(\mathbb{R})} \right.$$

$$\left. + \sum_{j\geq l} \lambda_j 2^{(j-l)\varepsilon q} \|f_{\widehat{p},j}\|^q_{L_q(\mathbb{R})} \right).$$

Changing the order of summation and roughly estimating $\lambda_j \leq 1$ gives

$$\|f\|^q_{B^s_q(L_q(\mathbb{R}))} \lesssim \sum_{j=-1}^{\infty} 2^{jsq} \|f_{\widehat{p},j}\|^q_{L_q(\mathbb{R})} \left(\sum_{l>j} 2^{-(l-j)(t-\varepsilon-s)q} (\widehat{p}+1)^{2tq} + \sum_{l\leq j} 2^{(j-l)(\varepsilon-s)q} \right)$$

$$\lesssim \sum_{j=-1}^{\infty} 2^{jsq} (\widehat{p}+1)^{2tq} \|f_{\widehat{p},j}\|^q_{L_q(\mathbb{R})},$$

where due to the choice of ε the geometric series converge. Combining this result with the upper estimates (4.5.3), (4.5.4) we conclude

$$\|f\|^q_{B^s_q(L_q(\mathbb{R}))} \lesssim \sum_{j=-1}^{\infty} 2^{jsq} (\widehat{p}+1)^{2tq} (\widehat{p}+1)^{q-1} 2^{j(\frac{1}{2}-\frac{1}{q})q} \left(\sum_{p=0}^{\widehat{p}} \sum_{k\in\mathbb{Z}} |c_{p,j,k}|^q \right)$$

$$= (\widehat{p}+1)^{2tq+q-1} \left(\sum_{p=0}^{\widehat{p}} \sum_{k\in\mathbb{Z}} \sum_{j=-1}^{\infty} 2^{j(s+\frac{1}{2}-\frac{1}{q})q} |c_{p,j,k}|^q \right).$$

Since the representation was arbitrary, taking the infimum gives the claim. $\qquad \square$

The following theorem is the first main result of this section. It generalises the upper estimate in (4.5.1) to the quarklet case.

Theorem 4.27. *Define the full quarklet system by*

$$\Psi := \{\psi_{p,j,k} : p \geq 0, j \geq -1, k \in \mathbb{Z}\}. \tag{4.5.8}$$

Then, for fixed $\delta > 1$ and $s < t < m - 1 + \frac{1}{q}$ we have the following norm estimate:

$$\|f\|_{B^s_q(L_q(\mathbb{R}))} \lesssim \inf_{(4.5.10)} \left(\sum_{p\geq 0} \sum_{j\geq -1} \sum_{k\in\mathbb{Z}} (p+1)^{(2t+1-\frac{1}{q}+\delta-\frac{\delta}{q})q} 2^{j(s+\frac{1}{2}-\frac{1}{q})q} |c_{p,j,k}|^q \right)^{1/q},$$

$$\tag{4.5.9}$$

$$c : f = \sum_{p\geq 0} \sum_{j\geq -1} \sum_{k\in\mathbb{Z}} c_{p,j,k} \psi_{p,j,k}. \tag{4.5.10}$$

Proof. Consider the series

$$f = \sum_{p\geq 0} f_p = \sum_{p\geq 0} \sum_{j\geq -1} \sum_{k\in\mathbb{Z}} c_{p,j,k} \psi_{p,j,k}.$$

Using the triangle inequality and Hölder's inequality for series with $\frac{1}{q} + \frac{1}{q'} = 1$ it follows

$$\|f\|^q_{B^s_q(L_q(\mathbb{R}))} \leq \left(\sum_{p \geq 0} \|f_p\|_{B^s_q(L_q(\mathbb{R}))} \right)^q$$

$$= \left(\sum_{p \geq 0} (p+1)^{-\delta/q'} (p+1)^{\delta/q'} \|f_p\|_{B^s_q(L_q(\mathbb{R}))} \right)^q$$

$$\leq \left(\sum_{p \geq 0} (p+1)^{-\delta} \right)^{q/q'} \sum_{p \geq 0} (p+1)^{\delta q/q'} \|f_p\|^q_{B^s_q(L_q(\mathbb{R}))}.$$

With $C_\delta = \left(\sum_{p \geq 0} (p+1)^{-\delta} \right)^{q-1}$ and (4.5.6) applied to f_p we obtain

$$\|f\|^q_{B^s_q(L_q(\mathbb{R}))} \lesssim C_\delta \sum_{p \geq 0} (p+1)^{\delta(q-1)} (p+1)^{2tq+q-1} \sum_{j \geq -1} \sum_{k \in \mathbb{Z}} 2^{j(s+\frac{1}{2}-\frac{1}{q})q} |c_{p,j,k}|^q$$

$$\lesssim \sum_{p \geq 0} \sum_{j \geq -1} \sum_{k \in \mathbb{Z}} (p+1)^{(2t+1-\frac{1}{q}+\delta-\frac{\delta}{q})q} 2^{j(s+\frac{1}{2}-\frac{1}{q})q} |c_{p,j,k}|^q,$$

which proves the claim. $\qquad\square$

The following theorem provides a characterisation of Besov spaces in terms of quarklets for a more general range of parameters. The proof mimics the proofs of the preceding two theorems with more cumbersome notation.

Theorem 4.28. *Let the quarklet system Ψ be defined by (4.5.8). Then, for fixed $\delta > 1$ and $s < t < m - 1 + \frac{1}{q}$ we have the following norm estimate:*

$$\|f\|_{B^s_r(L_q(\mathbb{R}))} \lesssim \inf_{(4.5.12)} \left(\sum_{p \geq 0} \sum_{j \geq -1} (p+1)^{(2t+1-\frac{1}{q}+\delta-\frac{\delta}{r})r} 2^{j(s+\frac{1}{2}-\frac{1}{q})r} \left(\sum_{k \in \mathbb{Z}} |c_{p,j,k}|^q \right)^{r/q} \right)^{1/r},$$
$$(4.5.11)$$

$$c : f = \sum_{p \geq 0} \sum_{j \geq -1} \sum_{k \in \mathbb{Z}} c_{p,j,k} \psi_{p,j,k}. \qquad (4.5.12)$$

Proof. We perform the proof by similar steps as in the preceding proof. First consider the series

$$f = \sum_{j \geq -1} f_{\widehat{p},j} = \sum_{p=0}^{\widehat{p}} \sum_{j \geq -1} \sum_{k \in \mathbb{Z}} c_{p,j,k} \psi_{p,j,k}.$$

Again, we use definition (1.1.29) of the Besov norm. Applying Jensen's inequality and the Bernstein estimate (4.1.22), we obtain

$$\|f\|^r_{B^s_r(L_q(\mathbb{R}))} \lesssim \sum_{j=-1}^{\infty} 2^{jsr} (\widehat{p}+1)^{2tr} \|f_{\widehat{p},j}\|^r_{L_q(\mathbb{R})}.$$

From the upper estimates (4.5.3), (4.5.4) we conclude

$$\|f\|^r_{B^s_r(L_q(\mathbb{R}))} \lesssim \sum_{j=-1}^{\infty} 2^{jsr}(\widehat{p}+1)^{2tr}(\widehat{p}+1)^{(1-\frac{1}{q})r}2^{j(\frac{1}{2}-\frac{1}{q})r}\left(\sum_{p=0}^{\widehat{p}}\sum_{k\in\mathbb{Z}}|c_{p,j,k}|^q\right)^{r/q}.$$

We end up with the auxiliary result

$$\|f\|^r_{B^s_r(L_q(\mathbb{R}))} \lesssim (\widehat{p}+1)^{(2t+1-\frac{1}{q})r}\sum_{j=-1}^{\infty}2^{j(s+\frac{1}{2}-\frac{1}{q})r}\left(\sum_{p=0}^{\widehat{p}}\sum_{k\in\mathbb{Z}}|c_{p,j,k}|^q\right)^{r/q}. \qquad (4.5.13)$$

In the second step we apply this auxiliary result to the series

$$f = \sum_{p\geq 0} f_p = \sum_{p\geq 0}\sum_{j\geq -1}\sum_{k\in\mathbb{Z}}c_{p,j,k}\psi_{p,j,k}.$$

Following the lines of the preceding proof leads to

$$\|f\|^r_{B^s_r(L_q(\mathbb{R}))} \lesssim C_\delta \sum_{p\geq 0}(p+1)^{\delta(r-1)}\|f_p\|^r_{B^s_r(L_q(\mathbb{R}))}.$$

By definition of f_p it holds $c_{\widetilde{p},j,k} = 0$ for $\widetilde{p} \neq p$, in particular it follows $\widehat{p} = p$. Combining this and (4.5.13) gives

$$\|f\|^r_{B^s_r(L_q(\mathbb{R}))} \lesssim \sum_{p\geq 0}(p+1)^{\delta(r-1)}(p+1)^{(2t+1-\frac{1}{q})r}\sum_{j=-1}^{\infty}2^{j(s+\frac{1}{2}-\frac{1}{q})r}\left(\sum_{k\in\mathbb{Z}}|c_{p,j,k}|^q\right)^{r/q}$$

$$= \sum_{p\geq 0}\sum_{j=-1}^{\infty}(p+1)^{(2t+1-\frac{1}{q}+\delta-\frac{\delta}{r})r}2^{j(s+\frac{1}{2}-\frac{1}{q})r}\left(\sum_{k\in\mathbb{Z}}|c_{p,j,k}|^q\right)^{r/q}.$$

Since the representation of f was arbitrary, the claim follows. $\qquad\square$

4.6 Characterisation of Besov Spaces for $q < 1$

We transfer the stability results from the previous section to the case of quasi-Banach spaces. We proceed in an analogous way as before, with the usual modifications for quasi-norms. Note that we have the additional restriction $s > \frac{1}{q} - 1$, cf. [73]. We begin with auxiliary results.

Proposition 4.29. *Let $\widehat{p} \in \mathbb{N}_0$, $0 < q < 1$. Let the cardinal B-spline quark φ_p be defined by (4.1.1). Then the following upper estimate holds true for all sequences $c = \{c_{p,k}\}_{p=0,\ldots,\widehat{p},k\in\mathbb{Z}}$:*

$$\|\sum_{p=0}^{\widehat{p}}\sum_{k\in\mathbb{Z}}c_{p,k}\varphi_p(\cdot - k)\|_{L_q(\mathbb{R})} \lesssim \left(\sum_{p=0}^{\widehat{p}}\sum_{k\in\mathbb{Z}}|c_{p,k}|^q\right)^{1/q}. \qquad (4.6.1)$$

Proof. Instead of the triangle inequality in the case of Banach spaces we use the inequality $\|f + g\|_{L_q(\mathbb{R})}^q \leq \|f\|_{L_q(\mathbb{R})}^q + \|g\|_{L_q(\mathbb{R})}^q$ to immediately obtain

$$\| \sum_{p=0}^{\widehat{p}} \sum_{k\in\mathbb{Z}} c_{p,k} \varphi_p(\cdot - k) \|_{L_q(\mathbb{R})}^q \leq \sum_{p=0}^{\widehat{p}} \sum_{k\in\mathbb{Z}} |c_{p,k}|^q \|\varphi_p(\cdot - k)\|_{L_q(\mathbb{R})}^q$$

$$\lesssim \sum_{p=0}^{\widehat{p}} \sum_{k\in\mathbb{Z}} |c_{p,k}|^q.$$

$\square$

Proposition 4.30. *Let $\widehat{p} \in \mathbb{N}_0$, $0 < q < 1$ and let $j \geq 0$ be fixed. Let the quarklets $\psi_{p,j,k}$ be defined by (4.2.2). Then the following upper estimate holds true for all sequences $c = \{c_{p,j,k}\}_{p=0,\dots,\widehat{p},k\in\mathbb{Z}}$:*

$$\| \sum_{p=0}^{\widehat{p}} \sum_{k\in\mathbb{Z}} c_{p,j,k} \psi_{p,j,k} \|_{L_q(\mathbb{R})} \lesssim 2^{j(\frac{1}{2}-\frac{1}{q})} \left(\sum_{p=0}^{\widehat{p}} \sum_{k\in\mathbb{Z}} |c_{p,j,k}|^q \right)^{1/q}. \tag{4.6.2}$$

Proof. The proof is analogous to the proof of Proposition 4.29, in particular we use the transformation performed in the proof of Proposition 4.25. $\square$

Theorem 4.31. *Let the truncated quarklet system be defined by (4.5.5). Then, for $\frac{1}{q} - 1 < s < t < m$ we have the following norm estimate:*

$$\|f\|_{B_q^s(L_q(\mathbb{R}))} \lesssim \inf_{(4.6.4)} (\widehat{p}+1)^{\frac{2tm}{q(m-1+1/q)}} \left(\sum_{p=0}^{\widehat{p}} \sum_{j\geq-1} \sum_{k\in\mathbb{Z}} 2^{j(s+\frac{1}{2}-\frac{1}{q})q} |c_{p,j,k}|^q \right)^{1/q}, \tag{4.6.3}$$

$$c : f = \sum_{p=0}^{\widehat{p}} \sum_{j\geq-1} \sum_{k\in\mathbb{Z}} c_{p,j,k} \psi_{p,j,k}. \tag{4.6.4}$$

Proof. The proof is similar to the proof of Theorem 4.26, though with the usual modifications for the case of quasi-norms. We consider an arbitrary

$$f = \sum_{j\geq-1} f_{\widehat{p},j} = \sum_{p=0}^{\widehat{p}} \sum_{j\geq-1} \sum_{k\in\mathbb{Z}} c_{p,j,k} \psi_{p,j,k}.$$

We use definition (1.1.29) of the Besov quasi-norm: First we consider the modulus of smoothness. Using the inequality $\omega_m(f + g, h)_{L_q(\mathbb{R})}^q \leq \omega_m(f, h)_{L_q(\mathbb{R})}^q + \omega_m(g, h)_{L_q(\mathbb{R})}^q$, (1.1.29) for $\frac{1}{q} - 1 < s < t < m$, $j \leq l$ and $\omega_m(f, 2^{-l})_{L_q(\mathbb{R})}^q \leq 2^m \|f\|_{L_q(\mathbb{R})}^q$ for $j > l$ we get

$$\omega_m(f, 2^{-l})_{L_q(\mathbb{R})}^q \leq \sum_{j\leq-1} \omega_m(f_{\widehat{p},j}, 2^{-l})_{L_q(\mathbb{R})}^q$$

$$\lesssim \sum_{j<l} 2^{-ltq} \|f_{\widehat{p},j}\|_{B_q^t(L_q(\mathbb{R}))}^q + \sum_{j\geq l} \|f_{\widehat{p},j}\|_{L_q(\mathbb{R})}^q.$$

Since the functions $f_{\widehat{p},j}$ are contained in $V_{\widehat{p},j+1}$, with the Bernstein estimate (4.1.25) we obtain

$$\omega_m(f, 2^{-l})^q_{L_q(\mathbb{R})} \lesssim \sum_{j<l} 2^{-ltq}(\widehat{p}+1)^{\frac{2mtq}{q(m-1+1/q)}} 2^{jtq} \|f_{\widehat{p},j}\|^q_{L_q(\mathbb{R})} + \sum_{j\geq l} \|f_{\widehat{p},j}\|^q_{L_q(\mathbb{R})}$$

$$= \sum_{j<l} 2^{-(l-j)tq}(\widehat{p}+1)^{\frac{2mtq}{q(m-1+1/q)}} \|f_{\widehat{p},j}\|^q_{L_q(\mathbb{R})} + \sum_{j\geq l} \|f_{\widehat{p},j}\|^q_{L_q(\mathbb{R})}.$$

Combining this estimate with (1.1.29) leads to

$$\|f\|^q_{B^s_q(L_q(\mathbb{R}))} \lesssim \sum_{l=0}^{\infty} 2^{lsq} \sum_{j<l} 2^{-(l-j)tq}(\widehat{p}+1)^{\frac{2mtq}{q(m-1+1/q)}} \|f_{\widehat{p},j}\|^q_{L_q(\mathbb{R})} + \sum_{j\geq l} \|f_{\widehat{p},j}\|^q_{L_q(\mathbb{R})}$$

$$= \sum_{j=-1}^{\infty} \left((\widehat{p}+1)^{\frac{2mtq}{q(m-1+1/q)}} \sum_{l>j} 2^{lsq} 2^{-(l-j)tq} + \sum_{l\leq j} 2^{lsq} \right) \|f_{\widehat{p},j}\|^q_{L_q(\mathbb{R})}.$$

Estimating the geometric series and using the upper estimates (4.6.1), (4.6.2) yields

$$\|f\|^q_{B^s_q(L_q(\mathbb{R}))} \lesssim (\widehat{p}+1)^{\frac{2mtq}{q(m-1+1/q)}} \sum_{j=-1}^{\infty} 2^{jsq} \|f_{\widehat{p},j}\|^q_{L_q(\mathbb{R})}$$

$$\lesssim (\widehat{p}+1)^{\frac{2mtq}{q(m-1+1/q)}} \sum_{j=-1}^{\infty} 2^{jsq} 2^{j(\frac{1}{2}-\frac{1}{q})q} \left(\sum_{p=0}^{\widehat{p}} \sum_{k\in\mathbb{Z}} |c_{p,j,k}|^q \right)$$

$$= (\widehat{p}+1)^{\frac{2mtq}{q(m-1+1/q)}} \sum_{p=0}^{\widehat{p}} \sum_{j=-1}^{\infty} \sum_{k\in\mathbb{Z}} 2^{j(s+\frac{1}{2}-\frac{1}{q})q} |c_{p,j,k}|^q.$$

Since the representation was arbitrary, taking the infimum gives the claim. $\qquad\square$

The following Theorem states the main result of this section.

Theorem 4.32. *Let the full quarklet system be defined by (4.5.8). Then, for $\frac{1}{q} - 1 < s < t < m$ we have the following norm estimate:*

$$\|f\|_{B^s_q(L_q(\mathbb{R}))} \lesssim \inf_{(4.6.6)} \left(\sum_{p\geq 0} \sum_{j\geq -1} \sum_{k\in\mathbb{Z}} (p+1)^{\frac{2tmq}{q(m-1+1/q)}} 2^{j(s+\frac{1}{2}-\frac{1}{q})q} |c_{p,j,k}|^q \right)^{1/q}, \quad (4.6.5)$$

$$c: f = \sum_{p\geq 0} \sum_{j\geq -1} \sum_{k\in\mathbb{Z}} c_{p,j,k} \psi_{p,j,k}. \quad (4.6.6)$$

Proof. Consider the series

$$f = \sum_{p\geq 0} f_p = \sum_{p\geq 0} \sum_{j\geq -1} \sum_{k\in\mathbb{Z}} c_{p,j,k} \psi_{p,j,k}.$$

With (4.6.3) we immediately obtain

$$\|f\|^q_{B^s_q(L_q(\mathbb{R}))} \leq \sum_{p\geq 0} \|f\|^q_{B^s_q(L_q(\mathbb{R}))}$$

$$\lesssim \sum_{p\geq 0} (p+1)^{\frac{2tmq}{q(m-1+1/q)}} \sum_{j\geq -1} \sum_{k\in\mathbb{Z}} 2^{j(s+\frac{1}{2}-\frac{1}{q})q} |c_{p,j,k}|^q.$$

Since the representation was arbitrary, taking the infimum gives the claim. $\qquad\square$

4.7 Quarklet Frames for Sobolev Spaces

Since the Sobolev spaces H^s are the solution spaces for certain operator equations, we now focus on these spaces. Since they are Hilbert spaces and provide additional structure, we can formulate more precise results about quarklet systems in Sobolev spaces, in particular we have the frame property, cf. Section 1.3. For a direct proof of the frame property in Sobolev spaces we refer to [28, 57].

Theorem 4.33. *For* $0 \leq s < m - \frac{1}{2}$ *the weighted quarklet system*

$$
\begin{aligned}
\Psi^s_{\vec{\sigma}} &:= \{w_{p,j,s}\psi^{\vec{\sigma}}_{p,j,k} : (p,j,k) \in \nabla_{\vec{\sigma}}\}, \\
w_{p,j,s} &:= (p+1)^{-2s-\delta}2^{-sj}, \quad \delta > 1,
\end{aligned}
\tag{4.7.1}
$$

is a frame for $H^s_{\vec{\sigma}}(I)$.

Proof. The proof relies on an application of Proposition 1.20. Since the quarklet system contains an underlying Riesz basis, it suffices to show the lower inequality in (1.3.9). Applying Theorem 4.27 for the interval case with $q = 2$ and reweighting the quarklets immediately gives the claim. $\qquad\square$

Chapter 5

Quarklets on Domains

This chapter is dedicated to quarklet frames on domains, more precisely we construct quarklet frames on polygonal domains that can be decomposed into translated d-dimensional unit cubes. Domains with reentrant corners, such as the L-shaped domain, lead to singular solutions of elliptic partial differential equations. Therefore, they depict an important test case for adaptive schemes. In the case of wavelets, Riesz bases on domains have successfully been constructed by tensorising univariate wavelets and extending wavelets on cubes, cf. [13]. We transfer this ansatz to the case of quarklet frames. In Section 5.1 we describe how quarklet frames and even more general frames on cubes can be obtained from univariate frames. In Section 5.2 we describe the extension to the domains described above. Let us mention that the findings from this chapter have been published in [27] and the extended version [26].

5.1 Quarklets on Cubes

In this section we construct quarklet frames on cubes. We restrict the discussion to the case of the d-dimensional unit cube $\square := I^d$. We use a tensor product ansatz to derive frames based on the univariate quarklet frames from Chapter 4. This ansatz generalises the results from [44] from the case of Riesz bases to the case of frames. It turns out that in our case frames with underlying Riesz bases are necessary. Since the univariate quarklet frames contain a wavelet basis, this condition is fulfilled. In the case of Sobolev spaces, an additional difficulty comes into play, namely that the spaces $H^s_{\boldsymbol{\sigma}}(\square), \boldsymbol{\sigma} = (\vec{\sigma}_1, \ldots, \vec{\sigma}_d), \vec{\sigma}_i \in \{0, \lfloor s + \frac{1}{2} \rfloor\}$, are usually *not* of tensor product type. Fortunately, the following relations hold for $s \in [0, \infty) \setminus (\mathbb{N}_0 + \frac{1}{2})$, cf. [13]:

$$H^s_{\boldsymbol{\sigma}}(\square) := \bigcap_{i=1}^d H^s_i(\square), \tag{5.1.1}$$

where

$$H^s_i(\square) := L_2(I) \otimes \cdots \otimes L_2(I) \otimes H^s_{\vec{\sigma}_i}(I) \otimes L_2(I) \otimes \cdots \otimes L_2(I) \subset L_2(\square), \tag{5.1.2}$$

with $H^s_{\vec{\sigma}_i}(I)$ at the i-th spot. Therefore, we have to construct tensor quarklet frames for the spaces (5.1.2) and to check to which extent the frame property carries over to the intersection (5.1.1). Let us recall the definition of the tensor product of Hilbert spaces, cf. [50, Sec. 2]. The *free linear space* $F(\mathcal{H}^{(1)}, \mathcal{H}^{(2)})$ of two Hilbert spaces $\mathcal{H}^{(1)}, \mathcal{H}^{(2)}$ is defined as the family of *finite* linear combinations

$$F(\mathcal{H}^{(1)}, \mathcal{H}^{(2)}) := \left\{ \sum_{\lambda} a_{\lambda}(f_{\lambda}, g_{\lambda}) : a_{\lambda} \in \mathbb{R}, f_{\lambda} \in \mathcal{H}^{(1)}, g_{\lambda} \in \mathcal{H}^{(2)} \right\}. \tag{5.1.3}$$

The scalar product on $F(\mathcal{H}^{(1)}, \mathcal{H}^{(2)})$ is given by

$$\left\langle \sum_{\lambda} a_{\lambda}(f_{\lambda}, g_{\lambda}), \sum_{\mu} b_{\mu}(h_{\mu}, k_{\mu}) \right\rangle_{F(\mathcal{H}^{(1)}, \mathcal{H}^{(2)})} := \sum_{\lambda} \sum_{\mu} a_{\lambda} b_{\mu} \langle f_{\lambda}, h_{\mu} \rangle_{\mathcal{H}^{(1)}} \langle g_{\lambda}, k_{\mu} \rangle_{\mathcal{H}^{(2)}}. \tag{5.1.4}$$

We define an equivalence relation for elements $x, y \in F(\mathcal{H}^{(1)}, \mathcal{H}^{(2)})$ by

$$x \sim y :\Leftrightarrow x - y = \sum_{\lambda} \sum_{\mu} a_{\lambda} b_{\mu} (f_{\lambda}, g_{\mu}) - \left(\sum_{\lambda} a_{\lambda} f_{\lambda}, \sum_{\mu} b_{\mu} g_{\mu} \right),$$

$$a_{\lambda}, b_{\mu} \in \mathbb{R}, f_{\lambda} \in \mathcal{H}^{(1)}, g_{\mu} \in \mathcal{H}^{(2)}.$$

Then, the Hilbert space $\mathcal{H}^{(1)} \otimes \mathcal{H}^{(2)}$ is defined as the completion of the quotient space

$$F(\mathcal{H}^{(1)}, \mathcal{H}^{(2)})\big/_{\sim}. \tag{5.1.5}$$

We utilise the following theorem from [58] which states that tensor products of frames form frames for tensor products of general Hilbert spaces.

Theorem 5.1. *[58, Thm. 2.2] Let $\mathcal{H}^{(i)}, i \in \{1, \ldots, d\}$ be Hilbert spaces with frames $\mathcal{F}_{\mathcal{H}^{(i)}} = \{f_{i,\lambda_i}\}_{\lambda_i \in \mathcal{I}_i}$ with frame bounds A_i, B_i. Then, the system*

$$\left\{ \bigotimes_{i=1}^{d} f_{i,\lambda_i} : f_{i,\lambda_i} \in \mathcal{F}_{\mathcal{H}^{(i)}} \right\} \tag{5.1.6}$$

is a frame for $\bigotimes_{i=1}^{d} \mathcal{H}^{(i)}$ with frame bounds $\prod_{i=1}^{d} A_i, \prod_{i=1}^{d} B_i$.

An application of Theorem 5.1 provides us with tensor product frames for all the spaces $H_i^s(\square)$ defined in (5.1.2). It remains to check under which conditions these frames also give rise to suitable systems in the intersection space $H_{\sigma}^s(\square)$ in (5.1.1). More general, the intersection of Hilbert spaces $\mathcal{H}^{(i)}$ which all have to be contained in a Hilbert space $\mathcal{H}$ is defined as

$$\bigcap_{i=1}^{d} \mathcal{H}^{(i)} := \{f : \|f\|_{\bigcap_{i=1}^{d} \mathcal{H}^{(i)}} < \infty\}, \quad \|f\|_{\bigcap_{i=1}^{d} \mathcal{H}^{(i)}} := \left(\sum_{i=1}^{d} \|f\|_{\mathcal{H}^{(i)}}^2 \right)^{1/2}.$$

Quite surprisingly, to perform our proof, it is not sufficient that the individual system possesses the frame property. In addition, each of the frames must contain a Riesz basis. Although this assumption is in a certain sense restrictive, it is always satisfied since our quarkonial frames by construction contain a wavelet Riesz basis. The following lemma generalises Lemma 3.1.8 of [44] from the case of Riesz bases to the case of frames.

Lemma 5.2. *Let $\mathcal{F}_{\mathcal{H}} = \{f_\lambda\}_{\lambda \in \mathcal{I}}$ be a frame for a Hilbert space $\mathcal{H}$ such that for $i \in \{1, \ldots, d\}$ and some non-zero scalars $w_\lambda^{(i)}$, $\lambda \in \mathcal{I}$, the sets $\mathcal{F}_{\mathcal{H}^{(i)}} := \{(w_\lambda^{(i)})^{-1} f_\lambda\}_{\lambda \in \mathcal{I}}$ form frames for Hilbert spaces $\mathcal{H}^{(i)} \subset \mathcal{H}$. Furthermore we assume that there exists a Riesz basis $\mathcal{R}_{\mathcal{H}} := \{f_\lambda\}_{\lambda \in \mathcal{I}_{\mathcal{R}}} \subset \mathcal{F}_{\mathcal{H}}$ for $\mathcal{H}$ such that the sets $\mathcal{R}_{\mathcal{H}^{(i)}} := \{(w_\lambda^{(i)})^{-1} f_\lambda\}_{\lambda \in \mathcal{I}_{\mathcal{R}}}$ form Riesz bases for $\mathcal{H}^{(i)} \subset \mathcal{H}$. Then the collection*

$$\left\{ \left(\sum_{i=1}^{d} (w_\lambda^{(i)})^2 \right)^{-1/2} f_\lambda \right\}_{\lambda \in \mathcal{I}}$$

is a frame for $\bigcap_{i=1}^{d} \mathcal{H}^{(i)} \subset \mathcal{H}$.

Proof. It is sufficient to prove the lemma for the case $d = 2$. Then, the general result follows by induction. Let $f \in \mathcal{H}^{(1)} \cap \mathcal{H}^{(2)}$. Since $\mathcal{R}_{\mathcal{H}}$ is a Riesz basis for $\mathcal{H}$ we have a unique representation $f = \sum_{\lambda \in \mathcal{I}_{\mathcal{R}}} \hat{c}_\lambda f_\lambda$. Let B_i be the optimal upper frame bounds and $B_{\max} = \max\{B_1, B_2\}$. Then the frame property of $\mathcal{F}_{\mathcal{H}^{(i)}}$ in $\mathcal{H}^{(i)}$, $i \in \{1, 2\}$ implies

$$B_{\max}^{-1} \|f\|_{\mathcal{H}^{(i)}}^2 \leq B_i^{-1} \|f\|_{\mathcal{H}^{(i)}}^2 \leq \inf_{c^{(i)} \in \ell_2(\mathcal{I}):(c^{(i)})^T \mathcal{F}_{\mathcal{H}} = f} \sum_{\lambda \in \mathcal{I}} (w_\lambda^{(i)})^2 (c_\lambda^{(i)})^2 \qquad (5.1.7)$$

The definition of $\|\cdot\|_{\mathcal{H}^{(1)} \cap \mathcal{H}^{(2)}}$ and (5.1.7) lead to

$$B_{\max}^{-1} \|f\|_{\mathcal{H}^{(1)} \cap \mathcal{H}^{(2)}}^2 \leq \inf_{(c^{(1)}, c^{(2)}) \in \ell_2(\mathcal{I})^2 :(c^{(i)})^T \mathcal{F}_{\mathcal{H}} = f} \sum_{\lambda \in \mathcal{I}} (w_\lambda^{(1)})^2 (c_\lambda^{(1)})^2 + (w_\lambda^{(2)})^2 (c_\lambda^{(2)})^2$$

$$\leq \inf_{c \in \ell_2(\mathcal{I}):c^T \mathcal{F}_{\mathcal{H}} = f} \sum_{\lambda \in \mathcal{I}} \left((w_\lambda^{(1)})^2 + (w_\lambda^{(2)})^2 \right) c_\lambda^2,$$

$$(5.1.8)$$

showing the lower frame inequality. Let $A_i^{\mathcal{R}}$, $i \in \{1, 2\}$ be the optimal lower Riesz constants and $A_{\min}^{\mathcal{R}} = \min\{A_1^{\mathcal{R}}, A_2^{\mathcal{R}}\}$. For the upper frame inequality we use the unique representation and the Riesz basis properties of $\mathcal{R}_{\mathcal{H}^{(i)}}$ in $\mathcal{H}^{(i)}$, $i \in \{1, 2\}$ to estimate

$$\inf_{c \in \ell_2(\mathcal{I}):c^T \mathcal{F}_{\mathcal{H}} = f} \sum_{\lambda \in \mathcal{I}} \left((w_\lambda^{(1)})^2 + (w_\lambda^{(2)})^2 \right) c_\lambda^2 \leq \sum_{\lambda \in \mathcal{I}_{\mathcal{R}}} \left((w_\lambda^{(1)})^2 + (w_\lambda^{(2)})^2 \right) \hat{c}_\lambda^2$$

$$= \sum_{\lambda \in \mathcal{I}_{\mathcal{R}}} (w_\lambda^{(1)})^2 \hat{c}_\lambda^2 + \sum_{\lambda \in \mathcal{I}_{\mathcal{R}}} (w_\lambda^{(2)})^2 \hat{c}_\lambda^2$$

$$\leq (A_1^{\mathcal{R}})^{-1} \|f\|_{\mathcal{H}^{(1)}}^2 + (A_2^{\mathcal{R}})^{-1} \|f\|_{\mathcal{H}^{(2)}}^2$$

$$\leq (A_{\min}^{\mathcal{R}})^{-1} \|f\|_{\mathcal{H}^{(1)} \cap \mathcal{H}^{(2)}}^2,$$

$$(5.1.9)$$

Combining (5.1.8) and (5.1.9) proves the claim. $\qquad \square$

An application of Theorem 5.1 and Theorem 4.23 yields the following theorem, which is the first main result of this section.

Theorem 5.3. *Let* $\{\Psi_{\lambda_i}^{\vec{\sigma}_i}\}, i = 1, \ldots, d,$ *be a family of univariate boundary adapted quarklet frames of order* $m \geq 2$, *with* $\tilde{m}$ *vanishing moments,* $\tilde{m} \geq m$, *according to Theorem 4.23. Then the family*

$$\Psi_\sigma := \bigotimes_{i=1}^d \Psi_{\vec{\sigma}_i} = \left\{ (w_\lambda^{L_2})^{-1} \psi_\lambda^\sigma : \lambda \in \nabla_\sigma := \prod_{i=1}^d \nabla_{\vec{\sigma}_i} \right\}, \tag{5.1.10}$$

$$\psi_\lambda^\sigma := \bigotimes_{i=1}^d \psi_{\lambda_i}^{\vec{\sigma}_i}, \tag{5.1.11}$$

with the weights

$$w_\lambda^{L_2} := \prod_{i=1}^d (p_i + 1)^{\delta/2}, \quad \delta > 1, \tag{5.1.12}$$

is a quarkonial tensor frame for $L_2(\square)$.

By means of Theorem 5.1, Lemma 5.2 and Theorem 4.33 we also obtain quarkonial frames for the Sobolev space $H_\sigma^s(\square)$, which is a second main result.

Theorem 5.4. *Let* $\{\Psi_{\lambda_i}^{\vec{\sigma}_i}\}, i = 1, \ldots, d,$ *be a family of univariate boundary adapted quarklet frames of order* $m \geq 2$, *with* $\tilde{m}$ *vanishing moments,* $\tilde{m} \geq m$, *according to Theorem 4.23. Then the family*

$$\Psi_\sigma^s := \left\{ (w_\lambda^{H^s})^{-1} \psi_\lambda^\sigma : \lambda \in \nabla_\sigma \right\}, \tag{5.1.13}$$

with the weights

$$w_\lambda^{H^s} := \left(\sum_{i=1}^d (p_i + 1)^{4s+\delta_1} 4^{sj_i} \right)^{1/2} \prod_{i=1}^d (p_i + 1)^{\delta_2/2}, \quad \delta_1 > 1, \delta_1 + \delta_2 > 2, \tag{5.1.14}$$

is a frame for $H_\sigma^s(\square)$, $0 \leq s < m - \frac{1}{2}$, $s \notin \mathbb{N}_0 + \frac{1}{2}$.

5.2 Quarklets on Domains

As described above, we aim to construct quarklet frames on domains that can be decomposed into unit cubes. The course of the section is as follows: First we introduce the domains of interest as the union of parametric images of the unit cube and recall some ideas of [13] concerning extension operators and isomorphisms between Sobolev spaces on different domains. Next we describe in a general setting how a combination of frames on cubes, Bessel systems which include the image of an extension operator and simple extensions lead to frames for Sobolev spaces on our target domain $\Omega \subset \mathbb{R}^d$. Finally, we show that the general machinery can be applied to our setting and present explicit constructions. The main result of this section is presented in Theorem 5.15.

5.2.1 Preliminaries

We collect the basic tools which are needed to generalise Riesz bases on cubes to Riesz bases on general domains. Further information can be found in [13]. This approach can be used as a starting point of the new frame construction on general domains.

Let us first describe the types of domains we will be concerned with in the sequel. Let $\square := I^d$. Let $\{\square_0, \ldots, \square_N\}$ with $\square_j := \tau_j + \square$, $\tau_j \in \mathbb{Z}^d$, $j = 0, \ldots, N$ be a fixed finite set of hypercubes. We assume $\cup_{j=0}^{N} \square_j \subset \Omega \subset (\cup_{j=0}^{N} \overline{\square}_j)^{\mathrm{int}}$ and such that $\partial\Omega$ is the union of (closed) facets of the $\square_j$'s. Later on we will present a construction of frames for Sobolev spaces on Ω from frames for corresponding Sobolev spaces on the subdomains $\square_k$ by using extension operators. These extension operators form a crucial ingredient in the final construction, see again Subsection 5.2.3. The following conditions $(\mathcal{D}_1)$–$(\mathcal{D}_5)$ are taken from [13] and ensure the existence of suitable extension operators.

We set $\Omega_i^{(0)} := \square_i$, $i = 0, \ldots, N$ and create a sequence $(\{\Omega_i^{(q)} : q \leq i \leq N\})_{0 \leq q \leq N}$ of sets of polytopes, where each next entry in this sequence is created by joining two polytopes from the previous entry whose joint interface is part of a hyperplane. More precisely, we assume that for any $1 \leq q \leq N$, there exists a $q \leq \bar{i} = \bar{i}^{(q)} \leq N$ and $q - 1 \leq i_1 = i_1^{(q)} \neq i_2 = i_2^{(q)} \leq N$ such that

$(\mathcal{D}_1)$ $\Omega_{\bar{i}}^{(q)} = \left(\overline{\Omega_{i_1}^{(q-1)} \cup \Omega_{i_2}^{(q-1)}} \setminus \partial\Omega \right)^{\mathrm{int}}$ is connected, and the interface $J := \Omega_{\bar{i}}^{(q)} \setminus (\Omega_{i_1}^{(q-1)} \cup \Omega_{i_2}^{(q-1)})$ is part of a hyperplane,

$(\mathcal{D}_2)$ $\{\Omega_i^{(q)} : q \leq i \leq N, i \neq \bar{i}\} = \{\Omega_i^{(q-1)} : q - 1 \leq i \leq N, i \neq \{i_1, i_2\}\}$,

$(\mathcal{D}_3)$ $\Omega_N^{(N)} = \Omega$.

By construction, the boundary of each $\Omega_i^{(q)}$ is a union of facets of hypercubes $\square_j$. We define $\mathring{H}^s(\Omega_i^{(q)})$ to be the closure in $H^s(\Omega_i^{(q)})$ of the smooth functions that are supported in the interior of $\Omega_i^{(q)}$. In particular, homogeneous boundary conditions are imposed on those facets of $\square_j$ that lie inside $\partial\Omega$. Hence we have $\mathring{H}^s(\Omega_N^{(N)}) = H_0^s(\Omega)$ and for some $\boldsymbol{\sigma}(j) \in (\{0, \lfloor s + 1/2 \rfloor\}^2)^d$,

$$\mathring{H}^s(\Omega_j^{(0)}) = \mathring{H}^s(\square_j) = H_{\boldsymbol{\sigma}(j)}^s(\square_j).$$

The boundary conditions on the hypercubes that determine the spaces $\mathring{H}^s(\Omega_i^{(q)})$, and the order in which polytopes are joined should be chosen such that

$(\mathcal{D}_4)$ on the $\Omega_{i_1}^{(q-1)}$ and $\Omega_{i_2}^{(q-1)}$ sides of J, the boundary conditions are of order 0 and $\lfloor t + \frac{1}{2} \rfloor$, respectively,

and, w.l.o.g. assuming that $J = \{0\} \times \check{J}$ and $I \times \check{J} \subset \Omega_{i_1}^{(q-1)}$,

($\mathcal{D}_5$) for any function in $\mathring{H}^s(\Omega_{i_1}^{(q-1)})$ that vanishes near $\{0,1\} \times \check{J}$, its reflection in $\{0\} \times \mathbb{R}^{n-1}$ (extended with zero, and then restricted to $\Omega_{i_2}^{(q-1)}$) is in $\mathring{H}^s(\Omega_{i_2}^{(q-1)})$.

The condition ($\mathcal{D}_5$) can be formulated by saying that the order of the boundary condition at any subfacet of $\Omega_{i_1}^{(q-1)}$ adjacent to J should not be less than this order at its reflection in J, where in case this reflection is not part of $\partial\Omega_{i_2}^{(q-1)}$, the latter order should be read as the highest possible one $\lfloor s + \frac{1}{2} \rfloor$; and furthermore, that the order of the boundary condition at any subfacet of $\Omega_{i_2}^{(q-1)}$ adjacent to J should not be larger than this order at its reflection in J, where in case this reflection is not part of $\partial\Omega_{i_1}^{(q-1)}$, the latter order should be read as the lowest possible one 0.

Given $1 \le q \le N$, for $l \in \{1,2\}$, let $R_l^{(q)}$ be the *restriction* of functions on $\Omega_{\tilde{i}}^{(q)}$ to $\Omega_{i_l}^{(q-1)}$, let $\eta_2^{(q)}$ be the *extension* of functions on $\Omega_{i_2}^{(q-1)}$ to $\Omega_{\tilde{i}}^{(q)}$ by zero, and let $E_1^{(q)}$ be any *extension* that is well defined on Sobolev spaces on $\Omega_{i_1}^{(q-1)}$ to Sobolev spaces $\Omega_{\tilde{i}}^{(q)}$.

Roughly speaking, in every step of our construction we will glue together two adjacent domains. One ingredient in such a step will be a bijective operator between Sobolev spaces on those domains. In the following proposition, which is taken from [13, Prop. 2.1], we consider a more general framework and give conditions under which a class of mappings between a Banach space and the Cartesian product of two other Banach spaces consists of isomorphisms. In Proposition 5.6, cf. [13, Prop. 4.2], we apply these statements to our special case.

Proposition 5.5. *For normed linear spaces V and V_i ($i = 1, 2$), let $E_1 \in B(V_1, V)$, $\eta_2 \in B(V_2, V)$, $R_1 \in B(V, V_1)$, and $R_2 \in B(\mathrm{Ran}(\eta_2), V_2)$ be such that*

$$R_1 E_1 = \mathrm{Id}, \quad R_2 \eta_2 = \mathrm{Id}, \quad R_1 \eta_2 = 0, \quad \mathrm{Ran}(\mathrm{Id} - E_1 R_1) \subset \mathrm{Ran}(\eta_2).$$

Then

$$E = [E_1 \quad \eta_2] \in B(V_1 \times V_2, V) \text{ is boundedly invertible,}$$

with inverse

$$E^{-1} = \begin{bmatrix} R_1 \\ R_2(\mathrm{Id} - E_1 R_1) \end{bmatrix}.$$

Proposition 5.6. *Assume that $E_1^{(q)} \in B(\mathring{H}^s(\Omega_{i_1}^{(q-1)}), \mathring{H}^s(\Omega_{\tilde{i}}^{(q)}))$, $\eta_2^{(q)} \in B(\mathring{H}^s(\Omega_{i_2}^{(q-1)}), \mathring{H}^s(\Omega_{\tilde{i}}^{(q)}))$. Then,*

$$E^{(q)} := [E_1^{(q)} \quad \eta_2^{(q)}] \in B\Big(\prod_{l=1}^{2} \mathring{H}^s(\Omega_{i_l}^{(q-1)}), \mathring{H}^s(\Omega_{\tilde{i}}^{(q)}) \Big)$$

is boundedly invertible.

Proof. This is a direct application of Proposition 5.5 with $V_1 = \mathring{H}^s(\Omega_{i_1}^{(q-1)})$, $V_2 = \mathring{H}^s(\Omega_{i_2}^{(q-1)})$, $V = \mathring{H}^s(\Omega_{\tilde{i}}^{(q)})$, $E_1 = E_1^{(q)}$, $\eta_2 = \eta_2^{(q)}$ and $R_l = R_l^{(q)}$, for $l \in \{1, 2\}$. $\quad\square$

Sequential execution of extensions as in Proposition 5.6 induces an isomorphism from the Cartesian product of Sobolev spaces on the cubes $\square_j$ onto the Sobolev spaces on the target domain Ω.

Corollary 5.7. *For F being the composition for $q = 1, \ldots, N$ of the mappings $E^{(q)}$ from Proposition 5.6, trivially extended with identity operators in coordinates $i \in \{q-1, \ldots, N\} \setminus \{i_1^{(q)}, i_2^{(q)}\}$, it holds that*

$$F \in B\Big(\prod_{j=0}^{N} \mathring{H}^s(\square_j), H_0^s(\Omega) \Big). \tag{5.2.1}$$

is boundedly invertible.

Remark 5.8. If we apply F to Riesz bases on cubes $\square_j$ we end up with a Riesz basis on Ω. While this is also true for the case of frames, the way for the construction of frames in this paper will be a bit different, mainly to preserve the vanishing moments of the frames on cubes. Nevertheless, the operators $E^{(q)}$ as defined in Proposition 5.6 will play an important role in the construction process.

The next proposition provides the link between the extension approach and tensor products. It states that under the conditions $(\mathcal{D}_1)$–$(\mathcal{D}_5)$, the extensions $E_1^{(q)}$ can be constructed (essentially) as tensor products of *univariate extensions* with identity operators in the other Cartesian directions.

Proposition 5.9. *In the setting of $(\mathcal{D}_1)$, w.l.o.g. let $J = \{0\} \times \breve{J}$ and $I \times \breve{J} \subset \Omega_{i_1}^{(q-1)}$. Let G_1 be an extension operator of functions on I to functions on $(-1,1)$ such that*

$$G_1 \in B(L_2(I), L_2(-1,1)), \quad G_1 \in B(H^s(I), H_{(\lfloor s+\frac{1}{2} \rfloor, 0)}^s(-1,1)).$$

Let $T \in B(\mathring{H}^s(\Omega_{i_1}^{(q-1)}), \mathring{H}^s(\Omega_{i_2}^{(q-1)}))$ be defined as the composition of the restriction to $I \times \breve{J}$, followed by an application of

$$G_1 \otimes \mathrm{Id} \otimes \cdots \otimes \mathrm{Id},$$

an extension by 0 to $\Omega_{i_2}^{(q-1)} \setminus (-1,0) \times \breve{J}$ and a restriction to $\Omega_{i_2}^{(q-1)}$. Then, we define $E^{(q)} \in B(\prod_{l=1}^{2} \mathring{H}^s(\Omega_{i_l}^{(q-1)}), \mathring{H}^s(\Omega_{\breve{i}}^{(q)}))$ as the operator which is the identity operator if restricted to $\mathring{H}^s(\Omega_{i_1}^{(q-1)})$ and T if restricted to $\mathring{H}^s(\Omega_{i_2}^{(q-1)})$. By proceeding this way, $E^{(q)}$ is well-defined and boundedly invertible by Proposition 5.6.

5.2.2 Construction of Frames by Extension

Based on the setting outlined in Subsection 5.2.1, we will now describe a general procedure to construct frames for the Sobolev space $H^s(\Omega)$, provided that suitable frames and Riesz-bases, respectively, on the cubes $\square_j$ are given. Suitable frames and

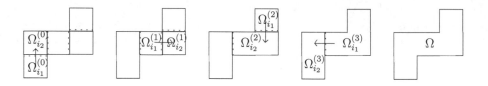

Figure 5.1: Example of a domain decomposition such that $\mathcal{D}_1 - \mathcal{D}_5$ are fulfilled. The arrows indicate the direction of the non-trivial extension. Dotted lines and solid lines indicate free and zero boundary conditions, respectively.

bases on cubes have been constructed in Section 5.1. Finally, a combination of the results of Subsection 5.2.2 and Section 5.1 will provide us with the desired quarklet frame, cf. Subsection 5.2.3.

For $j = 0, \ldots, N$, let $\boldsymbol{\Psi}_j$ be a frame for $L_2(\square_j)$, that renormalised in $H^s(\square_j)$, is a frame for $\overset{\circ}{H}{}^s(\square_j)$. Furthermore assume that there exists a Riesz basis $\boldsymbol{\Sigma}_j \subset \boldsymbol{\Psi}_j$ for $L_2(\square_j)$, that renormalised in $H^s(\square_j)$, is a Riesz basis for $\overset{\circ}{H}{}^s(\square_j)$. Renormalised versions of all sets will be indicated with an upper s. For $q = 0, \ldots, N$, $i = q, \ldots, N$ and $s \geq 0$ we define recursively

$$\boldsymbol{\Sigma}_i^{s,(q)} := \begin{cases} \boldsymbol{\Sigma}_i^s, & q = 0, \\ \boldsymbol{\Sigma}_{\hat{i}}^{s,(q-1)}, & 1 \leq q \leq N,\ i \neq \bar{i},\ \Omega_i^{(q)} = \Omega_{\hat{i}}^{(q-1)}, \\ E_1^{(q)}(\boldsymbol{\Sigma}_{i_1}^{s,(q-1)}) \cup \eta_2^{(q)}(\boldsymbol{\Sigma}_{i_2}^{s,(q-1)}), & 1 \leq q \leq N,\ i = \bar{i}. \end{cases} \quad (5.2.2)$$

We observe that $\boldsymbol{\Sigma}_N^{s,(N)}$ is exactly $F(\boldsymbol{\Sigma}_0^s, \ldots, \boldsymbol{\Sigma}_N^s)$, with F defined as in Corollary 5.7. Thus, it is a Riesz basis for $H_0^s(\Omega)$, cf. Proposition 1.22 (iii). For the frame construction, we have to assume the existence of an additional family $\boldsymbol{\Xi}^{s,(q)}$ which forms a Bessel system in $\overset{\circ}{H}{}^s(\Omega_i^{(q)})$, cf. (1.3.1), and satisfies $E_i^{(q)}(\boldsymbol{\Sigma}_i^{s,(q-1)}) \subset \boldsymbol{\Xi}^{s,(q)}$. Then for $q = 0, \ldots, N$, $i = q, \ldots, N$ and $s \geq 0$ we set

$$\boldsymbol{\Psi}_i^{s,(q)} := \begin{cases} \boldsymbol{\Psi}_i^s, & q = 0, \\ \boldsymbol{\Psi}_{\hat{i}}^{s,(q-1)}, & 1 \leq q \leq N,\ i \neq \bar{i},\ \Omega_i^{(q)} = \Omega_{\hat{i}}^{(q-1)}, \\ \boldsymbol{\Xi}^{s,(q)} \cup \eta_2^{(q)}(\boldsymbol{\Psi}_{i_2}^{s,(q-1)}), & 1 \leq q \leq N,\ i = \bar{i}. \end{cases} \quad (5.2.3)$$

The next proposition implies that, by proceeding this way, we indeed obtain suitable frames for $H_0^s(\Omega)$. Further information concerning the additional Bessel system as well as construction details can be found in Subsection 5.2.3, Remark 5.13.

Proposition 5.10. *For $q = 0, \ldots, N$, $i = q, \ldots, N$ and $s \geq 0$, let $\boldsymbol{\Psi}_i^{s,(q)}$ be defined as in (5.2.3). Then, $\boldsymbol{\Psi}^s := \boldsymbol{\Psi}_N^{s,(N)}$, is a frame for $H_0^s(\Omega)$.*

Proof. Let $1 \le q \le N$. Since $\boldsymbol{\Psi}_{i_2}^{s,(q-1)}$ is a Bessel system in $\mathring{H}^s(\Omega_{i_2}^{(q-1)})$ and $\eta_2^{(q)} \in B(\mathring{H}^s(\Omega_{i_2}^{(q-1)}), \mathring{H}^s(\Omega_{\vec{i}}^{(q)}))$, we can conclude that $\eta_2^{(q)}(\boldsymbol{\Psi}_{i_2}^{s,(q-1)})$ is a Bessel system in $\mathring{H}^s(\Omega_{\vec{i}}^{(q)})$, cf. Proposition 1.22 (i). Hence, $\boldsymbol{\Psi}_{\vec{i}}^{s,(q)} = \boldsymbol{\Xi}^{s,(q)} \cup \eta_2^{(q)}(\boldsymbol{\Psi}_{i_2}^{s,(q-1)})$ is a union of two Bessel systems and therefore a Bessel system in $\mathring{H}^s(\Omega_{\vec{i}}^{(q)})$, cf. Proposition 1.21 (i).

Since $E_1^{(q)}(\boldsymbol{\Sigma}_{i_1}^{s,(q-1)}) \subset \boldsymbol{\Xi}^{s,(q)}$ and $\boldsymbol{\Sigma}_{i_2}^{s,(q-1)} \subset \boldsymbol{\Psi}_{i_2}^{s,(q-1)}$, we conclude that $\boldsymbol{\Sigma}_{\vec{i}}^{s,(q)} \subset \boldsymbol{\Psi}_{\vec{i}}^{s,(q)}$. For $0 \le i \le N$, $\boldsymbol{\Sigma}_i^{s,(0)}$ is a Riesz basis for $\mathring{H}^s(\Omega_i^{(0)})$. Furthermore $E^{(q)} = [E_1^{(q)} \ \ \eta_2^{(q)}] \in B\left(\prod_{l=1}^2 \mathring{H}^s(\Omega_{i_l}^{(q-1)}), \mathring{H}^s(\Omega_{\vec{i}}^{(q)}) \right)$ as defined in Proposition 5.6 is boundedly invertible. Thus, we can conclude inductively that $\boldsymbol{\Sigma}_{\vec{i}}^{s,(q)} = E^{(q)}(\boldsymbol{\Sigma}_{i_1}^{s,(q-1)}, \boldsymbol{\Sigma}_{i_2}^{s,(q-1)})$ is a Riesz basis for $\mathring{H}^s(\Omega_{\vec{i}}^{(q)})$, cf. Proposition 1.22 (iii) . Hence, $\boldsymbol{\Psi}_{\vec{i}}^{s,(q)}$ as a Bessel system which contains a Riesz basis is a frame for $\mathring{H}^s(\Omega_{\vec{i}}^{(q)})$, cf. Proposition 1.21 (iii). Especially $\boldsymbol{\Psi}^s = \boldsymbol{\Psi}_N^{s,(N)}$ is a frame for $H_0^s(\Omega) = \mathring{H}^s(\Omega_N^{(N)})$. $\qquad\square$

5.2.3 Quarklet Frames on General Domains

Once we have constructed quarkonial tensor frames for scales of Sobolev spaces on cubes, the next step is clearly the generalisation to arbitrary domains as described in Subsection 5.2.1. To this end, we want to apply the general machinery as outlined in Subsection 5.2.2. Then, two basic ingredients have to be provided: suitable extension operators $E_1^{(q)}$, cf. (5.2.2), and the additional Bessel systems $\boldsymbol{\Xi}^{(q)}$, cf. (5.2.3).

Construction of scale-dependent extension operators

For $\vec{\sigma} = (\sigma_l, \sigma_r) \in \{0, \lfloor s+1/2 \rfloor\}^2$, the index set $\nabla_{\vec{\sigma}}^R$, cf. (2.2.30), and with $\vec{0} := (0,0)$, the functions in the univariate wavelet Riesz basis $\Sigma_{\vec{\sigma}}$, cf. (2.2.31), and its dual Riesz basis $\tilde{\Sigma}_{\vec{\sigma}}$ satisfy the following technical properties, cf. [13, Section 2]:

$(\mathcal{W}_1)$ $|\langle \tilde{\psi}_\lambda^{\vec{\sigma}}, u \rangle_{L_2(\mathcal{I})}| \lesssim 2^{-jt} \|u\|_{H^t(\mathrm{supp}\,\tilde{\psi}^{\vec{\sigma}})}$ $(u \in H^t(\mathcal{I}) \cap H_{\vec{\sigma}}^s(\mathcal{I}), \ \lambda \in \nabla_{\vec{\sigma}}^R)$, for some $\mathbb{N} \ni t > s$,

$(\mathcal{W}_2)$ $1 > \rho := \sup_{\lambda \in \nabla_{\vec{\sigma}}^R} 2^j \max(\mathrm{diam}\,\mathrm{supp}\,\tilde{\psi}_\lambda^{\vec{\sigma}}, \mathrm{diam}\,\mathrm{supp}\,\psi_\lambda^{\vec{\sigma}})$
$\qquad\qquad\quad \approx \inf_{\lambda \in \nabla_{\vec{\sigma}}^R} 2^j \max(\mathrm{diam}\,\mathrm{supp}\,\tilde{\psi}_\lambda^{\vec{\sigma}}, \mathrm{diam}\,\mathrm{supp}\,\psi_\lambda^{\vec{\sigma}})$,

$(\mathcal{W}_3)$ $\sup_{i,k \in \mathbb{N}_0} \#\{\lambda \in \nabla_{\vec{\sigma}}^R : j = i \wedge [k2^{-i}, (k+1)2^{-i}] \cap (\mathrm{supp}\,\tilde{\psi}_\lambda^{\vec{\sigma}} \cup \mathrm{supp}\,\psi_\lambda^{\vec{\sigma}}) \ne \emptyset\} < \infty$.

$(\mathcal{W}_4)$ $V_i^{\vec{\sigma}} := \mathrm{span}\{\psi_\lambda^{\vec{\sigma}} : \lambda \in \nabla_{\vec{\sigma}}^R, j \le i\} = V_i^{\vec{0}} \cap H_{\vec{\sigma}}^s(\mathcal{I})$,

$(\mathcal{W}_5)$ $\nabla_{\vec{\sigma}}^R$ is the disjoint union of $\nabla_{\sigma_\ell}^{R,(\ell)}, \nabla^{R,(I)}, \nabla_{\sigma_r}^{R,(r)}$ such that

a) $\displaystyle\sup_{\lambda \in \nabla_{\sigma_\ell}^{R,(\ell)}, x \in \mathrm{supp}\,\psi_\lambda^{\vec{\sigma}}} 2^j |x| \lesssim \rho$, $\quad \displaystyle\sup_{\lambda \in \nabla_{\sigma_r}^{R,(r)}, x \in \mathrm{supp}\,\psi_\lambda^{\vec{\sigma}}} 2^j |1 - x| \lesssim \rho$,

b) for $\lambda \in \nabla^{R,(I)}$, $\psi_\lambda^{\vec{\sigma}} = \psi_\lambda^{\vec{0}}$, $\tilde{\psi}_\lambda^{\vec{\sigma}} = \tilde{\psi}_\lambda^{\vec{0}}$, and the extensions of $\psi_\lambda^{\vec{0}}$ and $\tilde{\psi}_\lambda^{\vec{0}}$ by zero are in $H^s(\mathbb{R})$ and $L_2(\mathbb{R})$, respectively.

$(\mathcal{W}_6)$ $\begin{cases} \mathrm{span}\{\psi_\lambda^{\vec{0}}(1 - \cdot) : \lambda \in \nabla^{R,(I)}, \, j = i\} = \mathrm{span}\{\psi_\lambda^{\vec{0}} : \lambda \in \nabla^{R,(I)}, \, j = i\}, \\ \mathrm{span}\{\psi_\lambda^{(\sigma_\ell, \sigma_r)}(1 - \cdot) : \lambda \in \nabla_{\sigma_\ell}^{R,(\ell)}, \, j = i\} = \mathrm{span}\{\psi_\lambda^{(\sigma_r, \sigma_\ell)} : \lambda \in \nabla_{\sigma_r}^{R,(r)}, \, j = i\}, \end{cases}$

$(\mathcal{W}_7)$ $\begin{cases} \psi_\lambda^{\vec{\sigma}}(2^l \cdot) \in \mathrm{span}\{\psi_\mu^{\vec{\sigma}} : \mu \in \nabla_{\sigma_\ell}^{R,(\ell)}\} \quad (l \in \mathbb{N}_0, \, \lambda \in \nabla_{\sigma_\ell}^{R,(\ell)}), \\ \psi_\lambda^{\vec{0}}(2^l \cdot) \in \mathrm{span}\{\psi_\mu^{\vec{0}} : \mu \in \nabla^{R,(I)}\} \quad (l \in \mathbb{N}_0, \, \lambda \in \nabla^{R,(I)}). \end{cases}$

Let us first consider the simple *reflection*

$$\begin{aligned} (\breve{G}_1 v)(x) &:= v(x) & x \in I \\ (\breve{G}_1 v)(-x) &:= v(x) & x \in I, \end{aligned} \tag{5.2.4}$$

for any $v \in L_2(I)$. Obviously, we have

$$\begin{aligned} \breve{G}_1 &\in B(L_2(I), L_2(-1, 1)) \\ \breve{G}_1 &\in B(H^s(I), H^s(-1, 1)), \end{aligned} \tag{5.2.5}$$

for $s < 3/2$.

Remark 5.11. The use of the reflection operator has certain advantages and drawbacks. On the one hand, the reflection preserves the vanishing moment properties of the underlying frame elements which is a central ingredient for compression estimates, see Section 6.2. Moreover, the reflection possesses a moderate operator norm. On the other hand, it is clear that the reflection idea only works for Sobolev spaces H^s, $s < 3/2$, i.e., the resulting numerical schemes are restricted to second order elliptic equations. This bottleneck could be clearly avoided by using, e.g., higher order Hestenes extension operators. However, in recent studies, it has turned out that the norm of a Hestenes extension operator grows fast with respect to its order parameter. Moreover, it is not a priori clear if the vanishing moments are preserved. For this reason, in this paper we stick with the simple reflection operator.

Let η_1 and η_2 denote the extensions by zero of functions on I or on $(-1, 0)$ to functions on $(-1, 1)$, with R_1 and R_2 denoting their adjoints. With a univariate extension $\breve{G}_1$ as in (5.2.4) at hand, the obvious approach is to define $E_1^{(q)}$ according to Proposition 5.9 with $G_1 = \breve{G}_1$. A problem with the choice $G_1 = \breve{G}_1$ is that generally it does *not* imply the desirable property $\mathrm{diam}(\mathrm{supp}\, G_1 u) \lesssim \mathrm{diam}(\mathrm{supp}\, u)$. Indeed, think of the application of the reflection to a function u with a small support that is not located near the interface.

To solve this and the corresponding problem for the adjoint extension, following [13] we will apply our construction using the modified, *scale-dependent* univariate extension operator

$$G_1 : u \mapsto \sum_{\lambda \in \nabla_0^{R,(\ell)}} \langle u, \tilde{\psi}_\lambda^{\vec{0}} \rangle_{L_2(\mathcal{I})} \breve{G}_1 \psi_\lambda^{\vec{0}} + \sum_{\lambda \in \nabla^{R,(I)} \cup \nabla_0^{R,(r)}} \langle u, \tilde{\psi}_\lambda^{\vec{0}} \rangle_{L_2(\mathcal{I})} \eta_1 \psi_\lambda^{\vec{0}}. \tag{5.2.6}$$

So this operator reflects only wavelets that are supported near the interface. A proof of the following proposition can be found in [13, Proposition 5.2].

Proposition 5.12. *For $\vec{\sigma} \in \{0, \lfloor s + \frac{1}{2} \rfloor\}^2$, the scale-dependent extension G_1 from (5.2.6) satisfies*

$$G_1 \psi_\mu^{\vec{\sigma}} = \begin{cases} \eta_1 \psi_\mu^{\vec{\sigma}} & \text{when } \mu \in \nabla^{R,(I)} \cup \nabla_{\sigma_r}^{R,(r)}, \\ \breve{G}_1 \psi_\mu^{\vec{\sigma}} & \text{when } \mu \in \nabla_{\sigma_\ell}^{R,(\ell)}. \end{cases} \tag{5.2.7}$$

The resulting adjoint extension $G_2 := (\mathrm{Id} - \eta_1 G_1^*) \eta_2$ satisfies

$$G_2(\tilde{\psi}_\mu^{\vec{\sigma}}(1 + \cdot)) = \begin{cases} \eta_2(\tilde{\psi}_\mu^{\vec{\sigma}}(1 + \cdot)) & \text{when } \mu \in \nabla^{R,(I)} \cup \nabla_{\sigma_\ell}^{R,(\ell)}, \\ \breve{G}_2(\tilde{\psi}_\mu^{\vec{\sigma}}(1 + \cdot)) & \text{when } \mu \in \nabla_{\sigma_r}^{R,(r)}. \end{cases} \tag{5.2.8}$$

We have $G_1 \in B(L_2(I), L_2(-1, 1))$, and $G_1 \in B(H^s(I), H_{(\lfloor s + \frac{1}{2} \rfloor, 0)}^s(-1, 1))$, for $s < 3/2$.

Finally, for $\mu \in \nabla_{\vec{\sigma}}$, it holds that

$$\mathrm{diam}(\mathrm{supp}\, \breve{G}_1 \psi_\mu^{\vec{\sigma}}) \lesssim \mathrm{diam}(\mathrm{supp}\, \psi_\mu^{\vec{\sigma}}),$$
$$\mathrm{diam}(\mathrm{supp}\, \breve{G}_2 \tilde{\psi}_\mu^{\vec{\sigma}}) \lesssim \mathrm{diam}(\mathrm{supp}\, \tilde{\psi}_\mu^{\vec{\sigma}}).$$

Remark 5.13. In general, it is not possible to divide the univariate quarklet sets in such parts that statements similar to (5.2.7) and (5.2.8) hold. This can be explained as follows: since the univariate wavelets build a Riesz basis for a Sobolev space on the unit interval, every quarklet can be decomposed into wavelet elements. For quarklets near the boundary, it is not guaranteed that the participating wavelets of these decomposition lie exclusively in $\nabla^{R,(I)} \cup \nabla_{\sigma_r}^{R,(r)}$ or in $\nabla_{\sigma_\ell}^{R,(\ell)}$. Thus, it could happen that one part of the decomposition will be reflected and another part will be extended by zero. This would destroy the vanishing moments of the extended quarklets. Moreover, the wavelet decompositions of the quarklets have to be computed for every single quarklet, which is possible in theory but in practice very time-consuming. This is the reason why we use another approach with Bessel systems, which was already introduced in Section 5.2.2 and will be carried out further in the next subsection.

The Bessel systems $\Xi^{s,(q)}$

For the univariate quarklet frame $\Psi_{\vec{\sigma}}$ we can specify a *non-canonical* dual frame, cf. (1.3.6), if we augment the dual Riesz basis of the univariate wavelet basis $\Sigma_{\vec{\sigma}}$, cf. (2.2.31), with zero functions:

$$\Theta_{\vec{\sigma}} := \{\theta_\lambda^{\vec{\sigma}} : \lambda \in \nabla_{\vec{\sigma}}\}, \quad \theta_\lambda^{\vec{\sigma}} := \tilde{\psi}_\lambda^{\vec{\sigma}}, \text{ for } \lambda \in \nabla_{\vec{\sigma}}^R, \quad \theta_\lambda^{\vec{\sigma}} :\equiv 0, \text{ for } \lambda \in \nabla_{\vec{\sigma}} \setminus \nabla_{\vec{\sigma}}^R. \tag{5.2.9}$$

It is obvious that $\Theta_{\vec{\sigma}}$ is a dual frame of $\Psi_{\vec{\sigma}}$, since

$$\sum_{\lambda \in \nabla_{\vec{\sigma}}} \langle f, \theta_\lambda^{\vec{\sigma}} \rangle_{L_2(I)} \psi_\lambda^{\vec{\sigma}} = \sum_{\lambda \in \nabla_{\vec{\sigma}}^R} \langle f, \tilde{\psi}_\lambda^{\vec{\sigma}} \rangle_{L_2(I)} \psi_\lambda^{\vec{\sigma}} = f \quad \text{for all } f \in L_2(I).$$

With this dual frame at hand, $(\mathcal{W}_1)$-$(\mathcal{W}_3)$ also hold true if we replace $\nabla_{\vec{\sigma}}^R$ with $\nabla_{\vec{\sigma}}$ and $\tilde{\psi}_\lambda^{\vec{\sigma}}$ with $\theta_\lambda^{\vec{\sigma}}$. Also, it is possible to construct $\nabla_{\sigma_\ell}^{(\ell)} \supset \nabla_{\sigma_\ell}^{R,(\ell)}$, $\nabla^{(I)} \supset \nabla^{R,(I)}$, $\nabla_{\sigma_r}^{(r)} \supset \nabla_{\sigma_r}^{R,(r)}$, such that $\nabla_{\vec{\sigma}} = \nabla_{\sigma_\ell} \,\dot{\cup}\, \nabla^{(I)} \,\dot{\cup}\, \nabla_{\sigma_r}$, and

(1) $\displaystyle \sup_{\lambda \in \nabla_{\sigma_\ell}^{(\ell)},\, x \in \mathrm{supp}\,\psi_\lambda^{\vec{\sigma}}} 2^j |x| \lesssim \rho, \qquad \sup_{\lambda \in \nabla_{\sigma_r}^{(r)},\, x \in \mathrm{supp}\,\psi_\lambda^{\vec{\sigma}}} 2^j |1 - x| \lesssim \rho,$

(2) for $\lambda \in \nabla^{(I)}$, $\psi_\lambda^{\vec{\sigma}} = \psi_\lambda^{\vec{0}}$, $\theta_\lambda^{\vec{\sigma}} = \theta_\lambda^{\vec{0}}$, and the extensions of $\psi_\lambda^{\vec{0}}$ and $\theta_\lambda^{\vec{0}}$ by zero are in $H^s(\mathbb{R})$ and $L_2(\mathbb{R})$, respectively,

cf. $(\mathcal{W}_5)$. For $q \in \{1, \ldots, N\}$, $s \geq 0$, we define $\boldsymbol{\Psi}_{i_1,\ell}^{s,(q-1)}$ as the subset of functions $f \in \boldsymbol{\Psi}_{i_1}^{s,(q-1)}$ with the following properties:

(i) the support of f intersected with $I \times \check{J}$ is not empty,

(ii) the cube of origin $\square_i$ of f lies in the neighbourhood of $\{0\} \times \check{J}$, i.e., for all $\varepsilon > 0$: $\mathrm{diam}(\square_i, \{0\} \times \check{J}) < \varepsilon$,

(iii) the first Cartesian index of f restricted to its cube of origin is contained in $\nabla_0^{(\ell)}$.

With $\boldsymbol{\Psi}_{i_1,r}^{s,(q-1)} := \boldsymbol{\Psi}_{i_1}^{s,(q-1)} \setminus \boldsymbol{\Psi}_{i_1,\ell}^{s,(q-1)}$ we denote the complementary subset. Now we are ready to define the sets $\boldsymbol{\Xi}^{s,(q)}$ from (5.2.3) as

$$\boldsymbol{\Xi}^{s,(q)} := \check{E}_1^{(q)}(\boldsymbol{\Psi}_{i_1,\ell}^{s,(q-1)}) \cup \eta_1^{(q)}(\boldsymbol{\Psi}_{i_1,r}^{s,(q-1)}), \tag{5.2.10}$$

where $\check{E}_1^{(q)}$, $q \in \{1, \ldots, N\}$, are the operators corresponding to the simple reflection $\check{G}_1$.

Proposition 5.14. *For $q \in \{1, \ldots, N\}$, the set $\boldsymbol{\Xi}^{0,(q)}$ defined in (5.2.10) is a Bessel system for $L_2(\Omega_{\vec{i}}^{(q)})$ and $\boldsymbol{\Xi}^{s,(q)}$ a Bessel system for $\mathring{H}^s(\Omega_{\vec{i}}^{(q)})$, $0 < s < 3/2$, $s \neq \frac{1}{2}$. Also, we have $E_1^{(q)}(\boldsymbol{\Sigma}_{i_1}^{s,(q-1)}) \subset \boldsymbol{\Xi}^{s,(q)}$.*

Proof. Both $\boldsymbol{\Psi}_{i_1,\ell}^{0,(q-1)}$ and $\boldsymbol{\Psi}_{i_1,r}^{0,(q-1)}$ are subsets of the frame $\boldsymbol{\Psi}_{i_1}^{0,(q-1)}$ for $L_2(\Omega_{i_1}^{(q-1)})$. Hence, they are Bessel systems for $L_2(\Omega_{i_1}^{(q-1)})$. Since both $\check{E}_1^{(q)}$ and $\eta_1^{(q)}$ are bounded operators from $L_2(\Omega_{i_1}^{(q-1)})$ to $L_2(\Omega_{\vec{i}}^{(q)})$, the images $\check{E}_1^{(q)}(\boldsymbol{\Psi}_{i_1,\ell}^{0,(q-1)})$ and $\eta_1^{(q)}(\boldsymbol{\Psi}_{i_1,r}^{0,(q-1)})$ are Bessel systems for $L_2(\Omega_{\vec{i}}^{(q)})$, cf. Proposition 1.22 (i). For the renormalised versions we have to take care of the boundary conditions and the smoothness of the functions. For $s < 3/2$, it is $\check{G}_1 \in B(H_{(0,\lfloor s+\frac{1}{2}\rfloor)}^s(I), H_0^s(-1,1))$. Since the first Cartesian component of $\boldsymbol{\Psi}_{i_1,\ell}^{s,(q-1)}$ is in $H_{(0,\lfloor s+\frac{1}{2}\rfloor)}^s(I)$ the image of $\boldsymbol{\Psi}_{i_1,\ell}^{s,(q-1)}$ under $\check{E}_1^{(q)}$ is bounded in $\mathring{H}^s(\Omega_{\vec{i}}^{(q)})$ and therefore a Bessel system in $\mathring{H}^s(\Omega_{\vec{i}}^{(q)})$, cf. Proposition 1.22 (i). For the zero extension part we have $\eta_1 \in B(H_{(\lfloor s+\frac{1}{2}\rfloor,0)}^s(I), H_{(\lfloor s+\frac{1}{2}\rfloor,0)}^s(-1,1))$. The first Cartesian component of $\boldsymbol{\Psi}_{i_1,r}^{s,(q-1)}$ is in $H_{(\lfloor s+\frac{1}{2}\rfloor,0)}^s(I)$ and therefore the image of $\boldsymbol{\Psi}_{i_1,r}^{s,(q-1)}$ under

$\eta_1^{(q)}$ is also a Bessel system for $\mathring{H}^s(\Omega_{\hat{i}}^{(q)})$. The relation $E_1^{(q)}(\Sigma_{i_1}^{s,(q-1)}) \subset \Xi^{s,(q)}$ follows directly from (5.2.7) and (5.2.10) and the way how the sets $\Psi_{i_1,\ell}^{s,(q-1)}$ and $\Psi_{i_1,r}^{s,(q-1)}$ are defined. □

It remains to choose the index sets $\nabla_{\sigma_\ell}^{R,(\ell)}, \nabla^{R,(I)}, \nabla_{\sigma_r}^{R,(r)}$ and $\nabla_{\sigma_\ell}^{(\ell)}, \nabla^{(I)}, \nabla_{\sigma_r}^{(r)}$, appropriately. Let us assume that $m \geq 3$. From [67] we deduce that the index sets for which either the primal or dual wavelets depend on the incorporated boundary conditions are

$$\nabla_{\sigma_\ell}^{R,(\ell)} = \{(0,j,k) \in \nabla_{\vec{\sigma}} : k \in \nabla_{j,\sigma_\ell}^{(\ell)}\}, \quad \nabla_{\sigma_r}^{(r)} = \{(0,j,k) \in \nabla_{\vec{\sigma}} : k \in \nabla_{j,\sigma_r}^{(r)}\},$$

with

$$\nabla_{j,\sigma_\ell}^{(\ell)} = \begin{cases} \{0,\ldots,\frac{m+\tilde{m}-4}{2}\}, & j \geq j_0, \\ \{-m+1+\operatorname{sgn}\sigma_l,\cdots,\tilde{m}-2\}, & j = j_0 - 1, \end{cases}$$

and

$$\nabla_{j,\sigma_r}^{(r)} = \begin{cases} \{2^j - \frac{m+\tilde{m}-2}{2},\ldots,2^j-1\}, & j \geq j_0, \\ \{2^j - m - \tilde{m} + 2,\ldots,2^j-1-\operatorname{sgn}\sigma_r\}, & j = j_0 - 1. \end{cases}$$

The quarklet index sets are

$$\nabla_{\sigma_\ell}^{(\ell)} = \{(p,j,k) \in \nabla_{\vec{\sigma}} : k \in \nabla_{p,j,\sigma_\ell}^{(\ell)}\}, \quad \nabla_{\sigma_r}^{(r)} = \{(p,j,k) \in \nabla_{\vec{\sigma}} : k \in \nabla_{p,j,\sigma_r}^{(r)}\}$$

with

$$\nabla_{p,j,\sigma_\ell}^{(\ell)} = \begin{cases} \nabla_{j,\sigma_\ell}^{(\ell)}, & p = 0, \\ \{0 + \operatorname{sgn}\sigma_l,\ldots,0\}, & p > 0, j \geq j_0, \\ \{-m+1+\operatorname{sgn}\sigma_l,\cdots,-m+1\}, & p > 0, j = j_0 - 1, \end{cases}$$

and

$$\nabla_{p,j,\sigma_r}^{(r)} = \begin{cases} \nabla_{j,\sigma_r}^{(r)}, & p = 0, \\ \{2^j - 1,\ldots,2^j-1-\operatorname{sgn}\sigma_r\} & p > 0, \end{cases} \quad j \geq j_0.$$

In order to identify individual quarklets from the collections constructed by the applications of the extension operators, we have to introduce some more notations. For $0 \leq q \leq N$, we set the index sets

$$\boldsymbol{\nabla}_i^{(0)} := \boldsymbol{\nabla}_{\sigma(i)} \times \{i\} \text{ and, for } q > 0,$$

$$\boldsymbol{\nabla}_i^{(q)} := \begin{cases} \boldsymbol{\nabla}_{i_1}^{(q-1)} \cup \boldsymbol{\nabla}_{i_2}^{(q-1)} & \text{if } i = \bar{i}, \\ \boldsymbol{\nabla}_{\hat{i}}^{(q-1)} & \text{if } i \in \{q,\ldots,N\} \setminus \{\bar{i}\} \text{ and } \Omega_i^{(q)} = \Omega_{\hat{i}}^{(q-1)}. \end{cases}$$

(5.2.11)

We define the quarklets on the domains $\Omega_i^{(q)}$ as

$$\boldsymbol{\psi}_{\lambda,i}^{(0,i)} := \boldsymbol{\psi}_\lambda^{\sigma(i)}(\cdot - \tau_i),$$

(5.2.12)

and, for $q > 0$,

$$
\psi_{\lambda,n}^{(q,i)} := \begin{cases} \begin{cases} \breve{E}_1^{(q)} \psi_{\lambda,n}^{(q-1,i_1)} & (\lambda,n) \in \boldsymbol{\nabla}_{i_1,\ell}^{(q-1)} \\ \eta_1^{(q)} \psi_{\lambda,n}^{(q-1,i_1)} & (\lambda,n) \in \boldsymbol{\nabla}_{i_1,r}^{(q-1)} \\ \eta_2^{(q)} \psi_{\lambda,n}^{(q-1,i_2)} & (\lambda,n) \in \boldsymbol{\nabla}_{i_2}^{(q-1)} \end{cases} & \text{if } i = \bar{i}, \\[4ex] \psi_{\lambda,n}^{(q-1,\hat{i})} & \text{if } i \in \{q,\dots,N\} \setminus \{\bar{i}\} \text{ and } \Omega_i^{(q)} = \Omega_{\hat{i}}^{(q-1)}, \end{cases}
\tag{5.2.13}
$$

The index $n \in \{0,\dots,N\}$ indicates the cube $\square_n$ where the quarklet stems from. The subsets $\boldsymbol{\nabla}_{i_1,\ell}^{(q-1)}$ and $\boldsymbol{\nabla}_{i_1,r}^{(q-1)}$ are defined according to $\boldsymbol{\Psi}_{i_1,\ell}^{s,(q-1)}$ and $\boldsymbol{\Psi}_{i_1,r}^{s,(q-1)}$. With this notations at hand we are now able to formulate the main theorem of this chapter.

Theorem 5.15. *Let $\Psi_{\vec{\sigma}}$ denote a quarklet system of order $m \geq 2$, $\tilde{m}$ vanishing moments, $\tilde{m} \geq m$ and $m + \tilde{m}$ even, as constructed in Theorem 4.23. Furthermore, let $\Omega \in \mathbb{R}^d$ be a bounded domain that can be decomposed into cubes $\square_i$, $i = 0,\dots,N$. If we choose weights $\boldsymbol{w}_\lambda^{H^s}$ as in (5.1.14), the system*

$$
\boldsymbol{\Psi}^s := \left\{ (\boldsymbol{w}_\lambda^{H^s})^{-1} \psi_\alpha : \boldsymbol{\alpha} = (\lambda,n) \in \boldsymbol{\nabla} \right\}, \quad \delta_1 > 1, \ \delta_1 + \delta_2 > 2,
\tag{5.2.14}
$$

with $\psi_\alpha := \psi_{\lambda,n}^{(N,N)}$, cf. (5.2.13), $\boldsymbol{\nabla} := \boldsymbol{\nabla}_N^{(N)}$, cf. (5.2.11), is a frame for $H_0^s(\Omega)$, $0 \leq s < \frac{3}{2}$, $s \neq \frac{1}{2}$.

Figures 5.2 and 5.3 show quarks and quarklets on the L-shaped domain. Those functions completely lying on one subcube arise only from tensorisation of univariate functions, whereas the functions lying on two subcubes are tensorised and reflected at the interface. For a detailed description of the decomposition of the L-shaped domain we refer to Section 8.1.

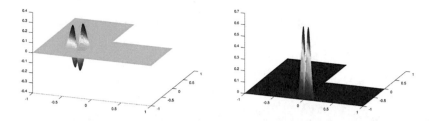

Figure 5.2: Quark generators of polynomial degree $p = (1,1)$ on the L-shaped domain. Left: Quark supported on one subcube. Right: Quark reflected at the interface with support in two cubes.

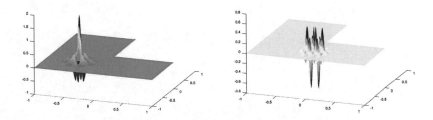

Figure 5.3: Quarklets of polynomial degree $p = (1, 1)$ on the L-shaped domain. Left: Quarklet supported on one subcube. Right: Quarklet reflected at the interface with support in two cubes.

Chapter 6

Adaptive Quarklet Schemes

In this chapter we discuss the application of quarklets in the numerical treatment of operator equations. Since suitably weighted quarklet systems form frames in Sobolev spaces, they can be utilised in generic frame schemes, cf. [20, 24, 25, 79, 81]. Those schemes rely on inexact applications of the biinfinite stiffness matrix. Accordingly, the employed frames have to fulfil certain conditions, in our case the vanishing moment property. The vanishing moments of the quarklets lead to *cancellation properties* and hence compressible stiffness matrices. First compression properties for univariate quarklets have been developed in [28], we concentrate on *second compression* in the univariate case in Section 6.2 and compression in the multivariate case in Section 6.3. The classical compression approach considers just differences in scales j and polynomial degrees p, on the other hand second compression also respects the translation parameter k and leads to economically beneficial cut-off rules. Our technique is inspired by second compression in the wavelet case, cf. [34, 74, 80]. We restrict our discussion to the Laplace operator, but generalisations are possible, cf. [32].

6.1 Discretisation of Operator Equations

This section is dedicated to the discretisation of elliptic operator equations. We proceed as described in [79]. In general, let $\mathcal{H}$ be a Hilbert space, then we consider the equation

$$Lu = f, \tag{6.1.1}$$

where $L \in \mathcal{L}(\mathcal{H}, \mathcal{H}')$ and $f \in \mathcal{H}'$. L is called *elliptic* if

$$\|Lu\|_{\mathcal{H}'} \sim \|u\|_{\mathcal{H}}, \quad u \in \mathcal{H}. \tag{6.1.2}$$

The ellipticity of L immediately implies that (6.1.1) has a unique solution u. With a given frame $\mathcal{F} := \{f_i\}_{i \in I}$, where I is a countable index set, (6.1.1) can be *discretised* in the following way. It holds $(Lu)(f_i) = f(f_i)$ for all $i \in I$. From $u \in \mathcal{H}$ and the stability of $\mathcal{F}$ we infer that u has a decomposition $u = \sum_{j \in I} c_j f_j$ with $c \in \ell_2(I)$. With the linearity of L we obtain the system of infinitely many equations

$$\sum_{j \in I} c_j (Lf_j)(f_i) = f(f_i), \quad i \in I. \tag{6.1.3}$$

Defining the *biinfinite stiffness matrix* $\boldsymbol{A}$ by

$$\boldsymbol{A}_{i,j} := (Lf_j)(f_i), \quad i,j \in I, \tag{6.1.4}$$

one interprets $\boldsymbol{A}$ as a bounded operator

$$\boldsymbol{A} : \ell_2(I) \to \ell_2(I)$$
$$\boldsymbol{c} \mapsto \{\sum_{j \in I} \boldsymbol{A}_{i,j} c_j\}_{i \in I}. \tag{6.1.5}$$

Analogously, with the definition of the *right-hand-side*

$$\boldsymbol{f} := \{f(f_i)\}_{i \in I}, \tag{6.1.6}$$

equation (6.1.3) is equivalent to the matrix equation

$$\boldsymbol{A}\boldsymbol{c} = \boldsymbol{f}. \tag{6.1.7}$$

In terms of the operators associated with the frame $\mathcal{F}$, see Section 1.3, we have

$$\boldsymbol{A} = T_{\mathcal{F}}^* L T_{\mathcal{F}}, \tag{6.1.8}$$
$$\boldsymbol{f} = T_{\mathcal{F}}^* f. \tag{6.1.9}$$

Now (6.1.7) is theoretically solvable with an iterative scheme, e.g., the damped Richardson iteration

$$\boldsymbol{c}^{(k+1)} := \boldsymbol{c}^{(k)} + \omega(\boldsymbol{f} - \boldsymbol{A}\boldsymbol{c}^{(k)}), \quad k = 0, 1, ..., \tag{6.1.10}$$

with suitable relaxation parameter $\omega < \frac{2}{|||\boldsymbol{A}|||}$. Of course, to implement such a scheme, further conditions have to be fulfilled. In particular, the right-hand-side $\boldsymbol{f}$ and the matrix vector product $\boldsymbol{A}\boldsymbol{c}^{(k)}$ have to be approximated by finitely supported sequences. Here we assume that the following routines are applicable for $\boldsymbol{A}$ and $\boldsymbol{f}$:

RHS$[\varepsilon, \boldsymbol{f}] \to \boldsymbol{f}_\varepsilon$: Computes for given $\boldsymbol{f} \in \ell_2(I)$, $\varepsilon > 0$ a finitely supported $\boldsymbol{f}_\varepsilon \in \ell_2(I)$ such that

$$\|\boldsymbol{f} - \boldsymbol{f}_\varepsilon\|_{\ell_2(I)} \leq \varepsilon. \tag{6.1.11}$$

APPLY$[\varepsilon, \boldsymbol{A}, \boldsymbol{c}] \to \boldsymbol{d}_\varepsilon$: Computes for given $\boldsymbol{A} \in \mathcal{L}(\ell_2(I))$, $\varepsilon > 0$ and finitely supported $\boldsymbol{c} \in \ell_2(I)$ a finitely supported $\boldsymbol{d}_\varepsilon \in \ell_2(I)$ such that

$$\|\boldsymbol{A}\boldsymbol{c} - \boldsymbol{d}_\varepsilon\|_{\ell_2(I)} \leq \varepsilon. \tag{6.1.12}$$

COARSE$[\varepsilon, \boldsymbol{c}] \to \boldsymbol{c}_\varepsilon$: Computes for given finitely supported $\boldsymbol{c} \in \ell_2(I)$, $\varepsilon > 0$ a finitely supported $\boldsymbol{c}_\varepsilon \in \ell_2(I)$ with at most N entries. Here, N is allowed to be a constant multiple of N_ε, where N_ε is the smallest number of entries such that it holds

$$\|\boldsymbol{c} - \boldsymbol{c}_\varepsilon\|_{\ell_2(I)} \leq \varepsilon. \tag{6.1.13}$$

The following realisation of the **APPLY** routine can be found in [79, Sec 3.2].
APPLY$[\varepsilon, \boldsymbol{A}, \boldsymbol{c}] \to \boldsymbol{d}_\varepsilon$:

- $q := \lceil \log((\# \operatorname{supp} \boldsymbol{c})^{1/2} \|\boldsymbol{c}\|_{\ell_2} \|\|\boldsymbol{A}\|\|\frac{2}{\varepsilon}) \rceil$.

- *Divide the elements of $\boldsymbol{c}$ into sets $\boldsymbol{C}_0, \dots, \boldsymbol{C}_q$, where for $0 \le i \le q - 1$, $\boldsymbol{C}_i$ contains the elements with modulus in $(2^{-i-1}\|\boldsymbol{c}\|_{\ell_2}, 2^{-i}\|\|_{\ell_2}]$, and possible remaining elements are put into $\boldsymbol{C}_q$.*

- *For $k = 0, 1, \dots$, generate vectors $\boldsymbol{c}_{[k]}$ by subsequently extracting $2^k - \lfloor 2^{k-1} \rfloor$ elements from $\cup_i \boldsymbol{C}_i$, starting from $\boldsymbol{C}_0$ and when it is empty continuing with $\boldsymbol{C}_1$ and so forth, until for some $k = l$ either $\cup_i \boldsymbol{C}_i$ becomes empty or*

$$\|\|\boldsymbol{A}\|\| \left\| \boldsymbol{c} - \sum_{k=0}^{l} \boldsymbol{c}_{[k]} \right\|_{\ell_2} \le \frac{\varepsilon}{2}. \tag{6.1.14}$$

In both cases $\boldsymbol{c}_{[l]}$ may contain less than $2^l - \lfloor 2^{l-1} \rfloor$ elements.

- *Compute the smallest $j \ge l$ such that*

$$\sum_{k_0}^{l} C_{j-k} \|\boldsymbol{c}_{[k]}\|_{\ell_2} \le \frac{\varepsilon}{2}. \tag{6.1.15}$$

- *For $0 \le k \le l$, compute the nonzero entries in the matrices $\boldsymbol{A}_{j-k}$ which have a column index in common with one of the entries of $\boldsymbol{c}_{[k]}$, and compute*

$$\boldsymbol{d}_\varepsilon := \sum_{k=0}^{l} \boldsymbol{A}_{j-k} \boldsymbol{c}_{[k]}. \tag{6.1.16}$$

For further reading on the above routines, in particular for conditions on the employed frame and its dual, we refer to [57, 79]. To investigate the practicability of such **APPLY** routines, the concept of *compressible* matrices is utilised.

Definition 6.1. Let $\boldsymbol{A} \in \mathcal{L}(\ell_2(I))$ and $s^* > 0$. $\boldsymbol{A}$ is called s^*-*compressible*, if for all $J \in \mathbb{N}$ there exist constants $C_J > 0$ and matrices $\boldsymbol{A}_J \in \mathcal{L}(\ell_2(I))$ which consist of at most $C_J 2^J$ non-trivial entries per row and column and fulfil

$$\sum_{J \in \mathbb{N}} 2^{Js} \|\boldsymbol{A} - \boldsymbol{A}_J\|_{\mathcal{L}(\ell_2(I))} < \infty \quad \text{for all } 0 \le s \le s^*. \tag{6.1.17}$$

With such a compressible matrix, the **APPLY** method is implementable, cf. [79]. The Schur lemma is a standard tool to prove compressibility for operators. We will employ it in the sequel of this chapter.

Lemma 6.2. *Let $A = (A_{i,j})_{i,j \in I}$ be an operator on $\ell_2(I)$ and $w_i > 0$ for all $i \in I$. If for $C > 0$ it holds*

$$\sup_{i \in I} w_i^{-1} \sum_{j \in I} |A_{i,j}| w_j \leq C,$$

$$\sup_{j \in I} w_j^{-1} \sum_{i \in I} |A_{i,j}| w_i \leq C,$$

then A is bounded with $\|A\|_{\mathcal{L}(\ell_2(I))} \leq C$.

6.2 Compression Techniques in the Univariate Case

In the following we precise our assumptions on the operator under consideration and the function system used for discretisation. For simplicity, we restrict our discussion to the Laplace operator, that means the operator L in (6.1.1) is given by the bilinear form

$$a(u,v) := \langle u', v' \rangle_{L_2(I)} = \int_0^1 u'(x) v'(x) \ \mathrm{d}x. \tag{6.2.1}$$

For discretisation, we use the $H^1(I)$ quarklet frame Ψ^1 defined in (4.7.1) with index set $\nabla = \nabla_{(0,0)}$ from (4.4.6). Note that this definition incorporates weights

$$w_\lambda = (p+1)^{-2-\delta} 2^{-j}. \tag{6.2.2}$$

Hence, the biinfinite stiffness matrix A is given by

$$A_{\lambda,\lambda'} = a(w_\lambda \psi_\lambda, w_{\lambda'} \psi_{\lambda'}), \quad \lambda, \lambda' \in \nabla. \tag{6.2.3}$$

A certain off-diagonal decay of the bilinear form is crucial for matrix compression. In [28] the following decay estimate has been proven for quarklets on the real line. It can be adapted to the case of quarklets on the interval since they possess the same amount of vanishing moments.

Proposition 6.3. *[28, Prop. 4] Let $\tilde{m} = m \geq 3$. For the unweighted quarklets it holds that*

$$|\langle \psi'_\lambda, \psi'_{\lambda'} \rangle_{L_2(I)}| \lesssim 2^{(j+j')} 2^{-|j-j'|(m-3/2)} (p+1)^{m-1} (p'+1)^{m-1} \quad \text{for all } \lambda, \lambda' \in \nabla. \tag{6.2.4}$$

The next step is to define the matrices A_J appropriately. This is done by *cut-off rules*. We cite a first compression result.

Theorem 6.4. *[28, Thm. 5] Let $J \in \mathbb{N}, a > 1, b \geq \frac{a}{a-1}$. Define the matrix A_J by*

$$(A_J)_{\lambda,\lambda'} = \begin{cases} A_{\lambda,\lambda'} & \text{if } a \log_2(1 + |p - p'|) + b|j - j'| \leq J, \\ 0 & \text{else.} \end{cases} \tag{6.2.5}$$

Then, $\boldsymbol{A}_J$ has asymptotically 2^J entries per row and column. Furthermore, for $\delta > (m-2)(\frac{a}{b}+1)$ the following estimate holds true:

$$||\boldsymbol{A}-\boldsymbol{A}_J||_{\mathcal{L}(\ell_2(\nabla))} \lesssim 2^{-J(m-2)/b}. \tag{6.2.6}$$

After recalling the basic concepts of matrix compression, we turn to a more involved approach. *Second compression* in the spirit of [80] makes use of the full amount of vanishing moments. In contrast, in [28, Prop. 4], this amount was restricted by the global regularity of the quarklets. To overcome this obstacle, we consider the local regularity. By definition, a quarklet ψ_λ is a piecewise polynomial of degree $p+m-1$. We subdivide the support of a quarklet into open intervals $\Xi_{\lambda,i}, i = 1, \ldots, n$, so that $\psi_\lambda|_{\Xi_{\lambda,i}}$ is a polynomial:

$$\psi_\lambda|_{\Xi_{\lambda,i}} \in \Pi_{p+m-1}, \tag{6.2.7}$$

$$\operatorname{supp}\psi_\lambda = \bigcup_{i=1}^{n} \overline{\Xi_{\lambda,i}}. \tag{6.2.8}$$

We define the *singular support* of a function f as the collection of all points for which f is not in $C^\infty(\mathbb{R})$. Due to the polynomial structure of the quarklets it holds

$$\operatorname{sing\ supp}\psi_\lambda = \operatorname{supp}\psi_\lambda \setminus \bigcup_{i=1}^{n}\Xi_{\lambda_i}. \tag{6.2.9}$$

For a fixed quarklet $\psi_{\lambda'}$ we define the index set of quarklets which do not intersect with the singular support of $\psi_{\lambda'}$:

$$A_{j,\lambda',i} := \{|\lambda| = j : \operatorname{supp}\psi_\lambda \subset \Xi_{\lambda',i}\}, \quad i = 1,\ldots,n, \tag{6.2.10}$$

Note that $j > j'$. In the sequel we improve some results from [28]. We start with a Bernstein estimate on patches $\Xi_{\lambda,i}$.

Proposition 6.5. *Let $\lambda \in \nabla$, $i \in \{1,\ldots,n\}$. For $1 \leq q \leq \infty$ it holds*

$$\|f^{(k)}\|_{L_q(\Xi_{\lambda,i})} \lesssim (p+1)^{2k}2^{jk}\|f\|_{L_q(\Xi_{\lambda,i})} \quad \textit{for all } k \in \mathbb{N}_0, f \in V_{p,j}, j \geq \lambda_j. \tag{6.2.11}$$

Proof. We use the fact that on each patch $\Xi_{\lambda,i}$ f is a polynomial of degree $p+m-1$. Applying Markovs inequality and estimating the size of $\Xi_{\lambda,i}$ by 2^{-j} leads to

$$\|f^{(k)}\|_{L_q(\Xi_{\lambda,i})} \lesssim \frac{(p+m-1)^{2k}}{|\Xi_{\lambda,i}|^k}\|f\|_{L_q(\Xi_{\lambda,i})}$$
$$\lesssim (p+1)^{2k}2^{jk}\|f\|_{L_q(\Xi_{\lambda,i})}.$$

$\square$

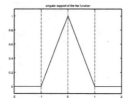 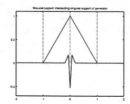 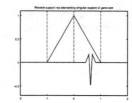

Figure 6.1: Blue: Hat function with indicated singular support (red). Black: Wavelet intersecting/not intersecting the singular support of the hat generator. Second compression heavily relies on distinguishing between these cases.

With these preparations we are ready to formulate a somewhat stronger version of Proposition 6.3.

Proposition 6.6. *Let λ' be fixed. Then, for $\lambda \in \bigcup_{j,i} A_{j,\lambda',i}$ and $t \in \{0,1,2,3\}$ it holds*

$$|\langle \psi'_\lambda, \psi'_{\lambda'} \rangle_{L_2(I)}| \lesssim 2^{(j+j')} 2^{-(j-j')(\tilde{m}-\frac{3}{2}+t)} (p'+1)^{2\tilde{m}-m-1+2t} (p+1)^{-(m-2)}. \quad (6.2.12)$$

Proof. The proof is similar to [28, Prop. 4]. Though, by the restriction $\lambda \in \bigcup_{j,i} A_{j,\lambda',i}$ the full amount of $\tilde{m}$ vanishing moments of the quarklet frame can be exploited. We start with using Hölder's inequality:

$$\langle \psi'_\lambda, \psi'_{\lambda'} \rangle_{L_2(I)} = \inf_{P \in \Pi_{\tilde{m}-3+t}} |\langle \psi'_\lambda, \psi'_{\lambda'} - P \rangle_{L_2(I)}|$$
$$\leq \inf_{P \in \Pi_{\tilde{m}-3+t}} \|\psi'_{\lambda'} - P\|_{L_\infty(\mathrm{supp}\,\psi_\lambda)} \|\psi'_\lambda\|_{L_1(\mathrm{supp}\,\psi_\lambda)}$$

Applying a Whitney type inequality and (4.1.21), (4.1.18) leads to

$$\langle \psi'_\lambda, \psi'_{\lambda'} \rangle_{L_2(I)} \lesssim |\mathrm{supp}\,\psi_\lambda|^{\tilde{m}-2+t} |\psi'_{\lambda'}|_{W^{\tilde{m}-2+t}_\infty(\mathrm{supp}\,\psi_\lambda)} \|\psi'_\lambda\|_{L_1(\mathrm{supp}\,\psi_\lambda)}$$
$$\lesssim 2^{-j(\tilde{m}-2+t)} (p+1)^{-(m-2)} 2^{j/2} \|\psi'^{(\tilde{m}-1+t)}_{\lambda'}\|_{L_\infty(\Xi_{\lambda',i})},$$

where $\mathrm{supp}\,\psi_\lambda \subset \Xi_{\lambda',i}$ for some i. We go on estimating with the Bernstein inequality (6.2.11):

$$\langle \psi'_\lambda, \psi'_{\lambda'} \rangle_{L_2(I)} \lesssim 2^{-j(\tilde{m}-2+t)} (p+1)^{-(m-2)} 2^{j/2} (p'+1)^{2(\tilde{m}-1+t)} 2^{j'(\tilde{m}-1+t)} \|\psi_{\lambda'}\|_{L_\infty(\Xi_{\lambda',i})}.$$

Again, using (4.1.18), we obtain for the latter part

$$\|\psi_{\lambda'}\|_{L_\infty(\Xi_{\lambda',i})} \leq \|\psi_{\lambda'}\|_{L_\infty(\mathbb{R})} \lesssim 2^{j'/2} (p'+1)^{-(m-1)}.$$

Putting all together finally proves the claim. $\square$

With the further assumptions $\delta > m - 3 + 2t$, $\widetilde{m} = m$ and the H^1-weights $w_\lambda = (p+1)^{-2-\delta}2^{-j}$ as in (6.2.2) we derive an estimate for the entries of the stiffness matrix for the Laplacian:

$$|a_{\lambda,\lambda'}| \lesssim 2^{-(j-j')(m-\frac{3}{2}+t)}(1 + |p' - p|)^{m-3+2t-\delta}, \quad \lambda' \in \nabla, \lambda \in \bigcup_{j,i} A_{j,\lambda',i}. \qquad (6.2.13)$$

The following theorem is the main result of this section. It states the effect of second compression in the case of quarklet frames on the interval.

Theorem 6.7. *Let $t \in \{0, 1, 2, 3\}$, $a > 0$, $b > 0$. With the notation*

$$d(\lambda, \lambda') := a \log_2(1 + |p' - p|) + b|j' - j|,$$

we define the cut-off rule for $J_1 > J_2 \in \mathbb{N}$:

$$(\mathbf{A}_J)_{\lambda,\lambda'} := \begin{cases} 0, & d(\lambda, \lambda') > J_1, \\ 0, & d(\lambda, \lambda') > J_2 \wedge \left(\lambda \in \bigcup_{j,i} A_{j,\lambda',i} \vee \lambda' \in \bigcup_{j',i} A_{j',\lambda,i}\right), \\ A_{\lambda,\lambda'}, & else. \end{cases}$$
$$(6.2.14)$$

Then, for $\mathbf{A}_J$ defined by (6.2.14) we have the following compression estimate:

$$\|\mathbf{A} - \mathbf{A}_J\|_{\mathcal{L}(\ell_2(\nabla))} \lesssim 2^{-(m-2)J_1/b} + 2^{-(m-2+t)J_2/b}, \qquad (6.2.15)$$

$$\delta > (m - 2 + t)(1 + \frac{a}{b}) + t. \qquad (6.2.16)$$

Furthermore, the number of non-zero entries in each row and column of $\mathbf{A}_J$ is of order 2^{J_1}.

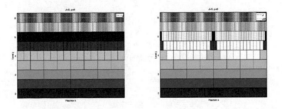

Figure 6.2: Visualisation of the cut-off rule for one column of $\mathbf{A}$. White blocks depict dropped entries. Second compression strategies (right) respect the translation parameter of the quarklets, classical compression strategies (left) do not.

Proof. Similarly as in [28] we employ Lemma 6.2, i.e., we show

$$\sup_{\lambda' \in \nabla} w_{\lambda'}^{-1} \sum_{\lambda \in \nabla} |(\boldsymbol{A} - \boldsymbol{A}_J)_{\lambda,\lambda'}| w_\lambda \lesssim 2^{-(m-2)J_1/b} + 2^{-(m-2+t)J_2/b}.$$

Let $\lambda' \in \nabla$ fixed. From Theorem 6.4 we already know

$$\sum_{\substack{\lambda \in \nabla: \\ d(\lambda,\lambda') > J_1}} w_{\lambda'}^{-1} |(\boldsymbol{A} - \boldsymbol{A}_J)_{\lambda,\lambda'}| w_\lambda \lesssim 2^{-(m-2)J_1/b},$$

hence we just have to deal with the case $J_2 < d(\lambda,\lambda') \leq J_1$ where second compression is possible. With $\beta_i := \lceil (J_i - a \log_2(1 + |p' - p|))/b \rceil, i = 1, 2$ we infer

$$\sum_{\substack{\lambda: J_2 < d(\lambda,\lambda') \leq J_1 \\ \lambda \in \bigcup_{j,i} A_{j,\lambda',i} \vee \lambda' \in \bigcup_{j',i} A_{j',\lambda,i}}} w_{\lambda'}^{-1} |(\boldsymbol{A} - \boldsymbol{A}_J)_{\lambda,\lambda'}| w_\lambda$$

$$\lesssim \sum_{\substack{\lambda: \\ a \log_2(1+|p'-p|) \leq J_1 \\ \beta_2 \leq |j'-j| < \beta_1 \\ \lambda \in \bigcup_{j,i} A_{j,\lambda',i} \vee \lambda' \in \bigcup_{j',i} A_{j',\lambda,i}}} w_{\lambda'}^{-1} |\boldsymbol{A}_{\lambda,\lambda'}| w_\lambda.$$

First let $j > j'$. With (6.2.13) and $w_\lambda = 2^{-j/2}$ we have

$$\alpha_1 := \sum_{\substack{p \geq 0, j > j': \\ a \log_2(1+|p'-p|) \leq J_1 \\ \beta_2 \leq j-j' < \beta_1 \\ \lambda \in \bigcup_{j,i} A_{j,\lambda',i}}} w_{\lambda'}^{-1} |\boldsymbol{A}_{\lambda,\lambda'}| w_\lambda$$

$$\lesssim \sum_{\substack{p \geq 0, j > j': \\ a \log_2(1+|p'-p|) \leq J_1 \\ \beta_2 \leq j-j' < \beta_1 \\ \lambda \in \bigcup_{j,i} A_{j,\lambda',i}}} 2^{j'/2} 2^{-(j-j')(m-\frac{3}{2}+t)} (1 + |p' - p|)^{m-3+2t-\delta} 2^{-j/2}$$

$$\lesssim \sum_{\substack{p \geq 0, j > j': \\ a \log_2(1+|p'-p|) \leq J_1 \\ \beta_2 \leq j-j' < \beta_1}} 2^{j-j'} 2^{j'/2} 2^{-(j-j')(m-\frac{3}{2}+t)} (1 + |p' - p|)^{m-3+2t-\delta} 2^{-j/2},$$

where for the last line we used that the number of wavelets with fixed level j intersecting with $\psi_{\lambda'}$ is of order $2^{j-j'}$. With an index transformation we obtain

$$\alpha_1 \lesssim \sum_{p: |p'-p| \leq 2^{J_1/a} - 1} \sum_{j=j'+\beta_2}^{j'+\beta_1} (1 + |p' - p|)^{m-3+2t-\delta} 2^{-(j-j')(m-2+t)}$$

$$= \sum_{p: |p'-p| \leq 2^{J_1/a} - 1} (1 + |p' - p|)^{m-3+2t-\delta} \sum_{j=0}^{\beta_1-\beta_2} 2^{-(j+\beta_2)(m-2+t)}$$

$$\lesssim \sum_{p: |p'-p| \leq 2^{J_1/a} - 1} (1 + |p' - p|)^{m-3+2t-\delta} 2^{-\beta_2(m-2+t)} \sum_{j=0}^{\infty} 2^{-j(m-2+t)}.$$

Estimating the geometric series by some constant leads to

$$\alpha_1 \lesssim \sum_{p:|p'-p|\leq 2^{J_1/a}-1} (1+|p'-p|)^{m-3+2t-\delta} 2^{-(J_2/b)(m-2+t)}(1+|p'-p|)^{\frac{a}{b}(m-2+t)}$$

$$= 2^{-(J_2/b)(m-2+t)} \sum_{p:|p'-p|\leq 2^{J_1/a}-1} (1+|p'-p|)^{m-3+2t-\delta+\frac{a}{b}(m-2+t)}.$$

With $\delta > (m-2+t)(1+\frac{a}{b})+t$ the latter sum can be estimated by twice the sum of a generalised harmonic series and for this case the claim follows. Now let $j < j'$. Using (6.2.13) with interchanged roles of j and j' and estimating the number of wavelets intersecting the support of $\psi_{\lambda'}$ by some constant we obtain

$$\alpha_2 := \sum_{\substack{p\geq 0, j<j': \\ a\log_2(1+|p'-p|)\leq J_1 \\ \beta_2\leq j'-j<\beta_1 \\ \lambda'\in\bigcup_{j',i} A_{j',\lambda,i}}} w_{\lambda'}^{-1}|\boldsymbol{A}_{\lambda,\lambda'}|w_\lambda$$

$$\lesssim \sum_{\substack{p\geq 0, j<j': \\ a\log_2(1+|p'-p|)\leq J_1 \\ \beta_2\leq j'-j<\beta_1 \\ \lambda'\in\bigcup_{j',i} A_{j',\lambda,i}}} 2^{j'/2}2^{-(j'-j)(m-\frac{3}{2}+t)}(1+|p'-p|)^{m-3+2t-\delta}2^{-j/2}$$

$$\lesssim \sum_{\substack{p\geq 0, j<j': \\ a\log_2(1+|p'-p|)\leq J_1 \\ \beta_2\leq j'-j<\beta_1}} 2^{j'/2}2^{-(j'-j)(m-\frac{3}{2}+t)}(1+|p'-p|)^{m-3+2t-\delta}2^{-j/2}.$$

Analogous calculations as in the first case lead to

$$\alpha_2 \lesssim \sum_{p:|p'-p|\leq 2^{J_1/a}-1} \sum_{j=j'-\beta_1}^{j'-\beta_2} (1+|p'-p|)^{m-3+2t-\delta}2^{-(j'-j)(m-2+t)}$$

$$= \sum_{p:|p'-p|\leq 2^{J_1/a}-1} (1+|p'-p|)^{m-3+2t-\delta} \sum_{j=j'-\beta_1}^{j'-\beta_2} 2^{(j-j')(m-2+t)}.$$

With $k = j - j'$ and $l = k - \beta_2$ we calculate

$$\alpha_2 \lesssim \sum_{p:|p'-p|\leq 2^{J_1/a}-1} (1+|p'-p|)^{m-3+2t-\delta} \sum_{k=-\beta_1}^{-\beta_2} 2^{k(m-2+t)}$$

$$= \sum_{p:|p'-p|\leq 2^{J_1/a}-1} (1+|p'-p|)^{m-3+2t-\delta} \sum_{k=\beta_2}^{\beta_1} 2^{-k(m-2+t)}$$

$$= \sum_{p:|p'-p|\leq 2^{J_1/a}-1} (1+|p'-p|)^{m-3+2t-\delta} \sum_{l=0}^{\beta_1-\beta_2} 2^{-(l+\beta_2)(m-2+t)}.$$

Estimating the geometric series by some constant and with the definition of β_2 we have

$$\alpha_2 \lesssim \sum_{p:|p'-p|\leq 2^{J_1/a}-1} (1+|p'-p|)^{m-3+2t-\delta} 2^{-\beta_2(m-2+t)} \sum_{l=0}^{\infty} 2^{-l(m-2+t)}$$

$$\lesssim \sum_{p:|p'-p|\leq 2^{J_1/a}-1} (1+|p'-p|)^{m-3+2t-\delta} 2^{-(J_2/b)(m-2+t)} (1+|p'-p|)^{\frac{a}{b}(m-2+t)}$$

$$= 2^{-(J_2/b)(m-2+t)} \sum_{p:|p'-p|\leq 2^{J_1/a}-1} (1+|p'-p|)^{m-3+2t-\delta+\frac{a}{b}(m-2+t)}$$

$$\lesssim 2^{-(J_2/b)(m-2+t)},$$

where again we need $\delta > (m-2+t)(1+\frac{a}{b})+t$. Putting all together gives the claim:

$$\sum_{\substack{\lambda:J_2<d(\lambda,\lambda')\leq J_1 \\ \lambda\in\bigcup_{j,i}A_{j,\lambda',i}\vee\lambda'\in\bigcup_{j',i}A_{j',\lambda,i}}} w_{\lambda'}^{-1}|(\boldsymbol{A}-\boldsymbol{A}_J)_{\lambda,\lambda'}|w_\lambda \lesssim \alpha_1+\alpha_2 \lesssim 2^{-(J_2/b)(m-2+t)}.$$

Next we consider the number of entries per row and column. The case $d(\lambda,\lambda') \leq J_2$ is covered by Theorem 6.4. Hence we just have to consider the case $J_2 < d(\lambda,\lambda') \leq J_1$ where (6.2.13) does not hold, i.e., for quarklets with intersecting singular support or small scale difference. Let λ' fixed and $j \geq j'$. Then the number of non-zero-entries is given by

$$N := \sum_{\substack{J_2<d(\lambda,\lambda')\leq J_1 \\ \lambda\notin\bigcup_{j,i}A_{j,\lambda',i}}} 1 \lesssim \sum_{p:a\log(1+|p-p'|)\leq J_1,} \sum_{j:\beta_2\leq|j-j'|<\beta_1,} \sum_{k:(p,j,k)\notin\bigcup_{j,i}A_{j,\lambda',i}} 1.$$

We additionally have to distinguish between coarse and far levels j. In our case, for $|j-j'| < \log_2(m)+1$ second compression ideas can never be applied. It holds $|\operatorname{supp}\psi_\lambda| \sim 2^{-j}2m$ and $|\Xi_{\lambda',i}| = 2^{-j'}$ due to the B-spline structure of the generators. To apply second compression ideas we need $|\operatorname{supp}\psi_\lambda| \leq |\Xi_{\lambda',i}|$ and hence $j-j' > \log_2(m)+1$. On coarse levels the number of intersecting quarklets is of order $2^{j-j'}$, on higher levels the number of quarklets intersecting the singular support of $\psi_{\lambda'}$ is of order 1. Accordingly splitting the sum leads to

$$N \lesssim \sum_{p:|p'-p|\leq 2^{J_1/a}-1} \sum_{j=j'+\beta_2}^{j'+\beta_1-1} 1 + \sum_{p:2^{J_2/a}m^{-b/a}<|p'-p|<2^{J_1/a}} \sum_{j=j'}^{j'+\log(m)+1} 2^{j-j'},$$

where the lower bound for p in the latter case is a consequence of $\log(m)+1 > \beta_2$. We go on estimating

$$N \lesssim \sum_{p:|p'-p|\leq 2^{J_1/a}-1} \frac{J_1-J_2}{b} + \sum_{p=p'+2^{J_2/a}m^{-b/a}-1}^{p'+2^{J_1/a}-1} \sum_{j=0}^{\log(m)+1} 2^j$$

$$\lesssim 2^{J_1/a}\frac{J_1-J_2}{b} + 2^{J_1/a} - 2^{J_2/a}m^{-b/a}.$$

Now let $j < j'$. In this case the number of intersecting quarklets is bounded by some constant. Thus we directly obtain

$$N \lesssim \sum_{p:|p'-p|\leq 2^{J_1/a-1}} \sum_{j=j'-\beta_1}^{j'-\beta_2} 1 \lesssim \sum_{p:|p'-p|\leq 2^{J_1/a-1}} \frac{J_1 - J_2}{b}$$

$$\lesssim 2^{J_1/a}\frac{J_1 - J_2}{b}.$$

$\square$

Corollary 6.8. *Choosing $J_2 = \frac{m-2}{m-2+t}J_1$ in (6.2.14), the matrix $\boldsymbol{A}$ is s^*-compressible with $s^* = \min\{a,b\}\frac{m-2}{b}$.*

Proof. Assuming a linear relation between the parameters J_1, J_2, i.e., $J_2 = cJ_1, c \in \mathbb{R}$, it follows from (6.2.15) that $\boldsymbol{A}$ is s^*-compressible with $s^* = \min\{a,b\}\min\{\frac{m-2}{b}, \frac{c(m-2+t)}{b}\}$. Equilibrating the arguments gives the claim. $\square$

Let us briefly discuss the gain of second compression in the univariate case. From Theorem 6.7 and Corollary 6.8 it follows that for $m = \widetilde{m} = 3$ we can choose $J_2 \in \{\frac{1}{2}J_1, \frac{1}{3}J_1, \frac{1}{4}J_1\}$ without loss of compressibility. A certain drawback is of course the stronger damping described in (6.2.16). Nevertheless second compression pays off, this finding will numerically be supported in Section 8.2.

6.3 Compression Techniques in the Multivariate Case

The techniques derived in Section 6.2 can, with some effort, be generalised to the multivariate case. The results concerning first compression have with some modifications been published in [27]. Let the domain Ω satisfy the assumptions of Chapter 5. For a fixed right-hand side $f \in H^{-1}(\Omega)$ we want to compute the solution $u \in H_0^1(\Omega)$ to (6.1.1), where

$$a(u,v) = \int_\Omega \nabla u \cdot \nabla v \, \mathrm{d}x = \sum_{k=1}^{d} \int_\Omega \frac{\partial u}{\partial x_k} \frac{\partial v}{\partial x_k} \, \mathrm{d}x. \tag{6.3.1}$$

In the setting of Chapter 5, the domain Ω is a hypercube or a union of finitely many translated copies thereof, and the frame elements $\boldsymbol{\psi}_\lambda$ are tensor products of univariate functions. Therefore, the individual entries $a(\boldsymbol{\psi}_\mu, \boldsymbol{\psi}_\lambda)$ of the stiffness matrix $\boldsymbol{A}$ are sums of products of univariate integrals. Let, for example, $\Omega = I^2$, and $\{f_\lambda : \lambda \in \mathcal{I}\}$ be a frame for $L_2(I)$ such that $\{w_\lambda^{-1}f_\lambda : \lambda \in \mathcal{I}\}$ is a frame for $H_0^1(I)$. Then,

$$\mathcal{F} := \left\{(w_{\lambda_1}^2 + w_{\lambda_2}^2)^{-1/2}f_{\lambda_1} \otimes f_{\lambda_2} : \lambda_1, \lambda_2 \in \mathcal{I}\right\}$$

is a frame for

$$H_0^1(I^2) = H_0^1(I) \otimes L_2(I) \cap L_2(I) \otimes H_0^1(I),$$

and the stiffness matrix $\mathbf{A}$ with respect to $\mathcal{F}$ is a sum of Kronecker products,

$$\mathbf{A} = \mathbf{D}_2^{-1}(\mathbf{B} \otimes \mathbf{G} + \mathbf{G} \otimes \mathbf{B})\mathbf{D}_2^{-1},$$

where $\mathbf{B} = (\int_0^1 f'_\lambda f'_\mu \, dx)_{\lambda,\mu \in \mathcal{I}}$ and $\mathbf{G} = (\int_0^1 f_\lambda f_\mu \, dx)_{\lambda,\mu \in \mathcal{I}}$ are one-dimensional stiffness and Gramian matrices, respectively and $\mathbf{D}_2 = (w_\lambda)_{\lambda \in \mathcal{I}^2}$, $\boldsymbol{\lambda} = (\lambda_1, \lambda_2)$, $w_\lambda := (w_{\lambda_1}^2 + w_{\lambda_2}^2)^{1/2}$. In the light of these tensor product techniques, we will utilise compression estimates for quarklet discretisations of one-dimensional elliptic equations. As we have seen in the two-dimensional case, the stiffness matrix of the Poisson equation (6.3.1) is a sum of Kronecker products of one-dimensional Laplacian and Gramian matrices. For $d \in \mathbb{N}$ dimensions this can be generalised easily to

$$\mathbf{A} = \mathbf{D}_d^{-1}(\mathbf{B} \otimes \mathbf{G} \otimes \ldots \otimes \mathbf{G} + \ldots + \mathbf{G} \otimes \ldots \otimes \mathbf{G} \otimes \mathbf{B})\mathbf{D}_d^{-1}.$$

Hence, to estimate the compressibility properties of the resulting stiffness matrix of the Laplacian (6.3.1), we need estimates for the inner products of the basic univariate quarks and quarklets. These preparations are made in the next propositions. Let again $\nabla = \nabla_{(0,0)}$ as in (4.4.6)

Proposition 6.9. *Let $m \geq 3$. There exists $C = C(m)$, such that the unweighted quarks and quarklets satisfy*

$$|\langle \psi_\lambda, \psi_{\lambda'} \rangle_{L_2(I)}| \lesssim ((p+1)(p'+1))^{m-1} 2^{-|j-j'|(m-1/2)} \qquad \text{for all } \lambda, \lambda' \in \nabla. \quad (6.3.2)$$

Proof. The combination of Lemma 4.21, (4.1.21), and for the last step (4.1.18) yields

$$\left|\langle \psi_\lambda, \psi_{\lambda'} \rangle_{L_2(I)}\right| \lesssim (p+1)^{-m} 2^{-j(m-\frac{1}{2})} |\psi_{\lambda'}|_{W_\infty^{m-1}(\operatorname{supp} \psi_\lambda^{\vec{\sigma}})}$$

$$\lesssim (p+1)^{-m} 2^{-j(m-\frac{1}{2})} (p'+1)^{2(m-1)} 2^{j'(m-1)} \|\psi_{\lambda'}\|_{L_\infty(I)}$$

$$= (p+1)^{-m} 2^{-j(m-\frac{1}{2})} (p'+1)^{2(m-1)} 2^{j'(m-\frac{1}{2})} \|\varphi_{p',0}\|_{L_\infty(I)}$$

$$\lesssim (p+1)^{-m} (p'+1)^{m-1} 2^{(j'-j)(m-\frac{1}{2})}.$$

The analogous result holds with interchanged roles of (p, j, k) and (p', j', k'). The minimum over both estimates yields (6.3.2). $\qquad \square$

We are able to formulate a stronger version for quarklets which do not intersect with the singular support of $\psi_{\lambda'}$:

Proposition 6.10. *Let $\lambda' \in \nabla$ be fixed. Then, for $\lambda \in \bigcup_{j,i} A_{j,\lambda',i}$ it holds*

$$|\langle \psi_\lambda, \psi_{\lambda'} \rangle_{L_2(I)}| \lesssim 2^{-(j-j')(\widetilde{m}+1/2)} (p'+1)^{2\widetilde{m}-m+1} (p+1)^{-m}. \quad (6.3.3)$$

Proof. The proof is analogous to the proof of Proposition 6.6. The support of ψ_λ does not intersect with the singular support of $\psi_{\lambda'}$, hence it holds $\psi_{\lambda'}|_{\operatorname{supp} \psi_\lambda} \in \Pi_{p+m-1}$. The quarklet $\psi_{\lambda'}$ has $\widetilde{m}$ vanishing moments, this allows us to conclude

$$|\langle \psi_\lambda, \psi_{\lambda'} \rangle_{L_2(I)}| = \inf_{P \in \Pi_{\widetilde{m}-1}} |\langle \psi_\lambda, \psi_{\lambda'} - P \rangle_{L_2(I)}|$$

$$\leq \inf_{P \in \Pi_{\widetilde{m}-1}} \|\psi_{\lambda'} - P\|_{L_\infty(\operatorname{supp} \psi_\lambda)} \|\psi_\lambda\|_{L_1(\operatorname{supp} \psi_\lambda)}.$$

Using a Whitney type inequality, the Bernstein estimates (4.1.21), (6.2.11), and the inequality (4.1.18) we get

$$|\langle \psi_\lambda, \psi_{\lambda'} \rangle_{L_2(I)}| \lesssim |\operatorname{supp} \psi_\lambda|^{\widetilde{m}} |\psi_{\lambda'}|_{W^{\widetilde{m}}_\infty(\operatorname{supp}\psi_\lambda)} \|\psi_\lambda\|_{L_1(\operatorname{supp}\psi_\lambda)}$$

$$\lesssim (2^{-j})^{\widetilde{m}}(p'+1)^{2\widetilde{m}}2^{j'\widetilde{m}}\|\psi_{\lambda'}\|_{L_\infty(\operatorname{supp}\psi_\lambda)}\|\psi_\lambda\|_{L_1(\operatorname{supp}\psi_\lambda)}$$

$$\lesssim 2^{-j\widetilde{m}}(p'+1)^{2\widetilde{m}}2^{j'\widetilde{m}}2^{j'/2}(p'+1)^{-(m-1)}2^{-j/2}(p+1)^{-m}$$

$$= 2^{-(j-j')(\widetilde{m}+1/2)}(p'+1)^{2\widetilde{m}-m+1}(p+1)^{-m}.$$

$\square$

We formulate an auxiliary lemma to improve the multivariate compression estimates from [26]. We estimate a series which decreases in a mixed geometric manner.

Lemma 6.11. *For $l \in \mathbb{N}$, $m \in \mathbb{R}$ and with $|\alpha| := \sum_{k=1}^d \alpha_k$ the following holds:*

$$\sum_{\substack{\alpha \in \mathbb{N}_0^d \\ |\alpha|=l}} 2^{-\sum_{k=1}^d \alpha_k(m-\delta_{k,d})} \lesssim 2^{-l(m-1)}. \tag{6.3.4}$$

Proof. We proceed by induction. Let $d = 2$. Splitting the sum leads to

$$\sum_{\substack{\alpha \in \mathbb{N}_0^2 \\ |\alpha|=l}} 2^{-\alpha_1 m - \alpha_2(m-1)} = \sum_{\substack{\alpha_1 \in \mathbb{N}_0 \\ \alpha_1 \leq l}} 2^{-\alpha_1 m} \sum_{\substack{\alpha_2 \in \mathbb{N}_0 \\ \alpha_2 = l-\alpha_1}} 2^{-\alpha_2(m-1)}$$

$$= \sum_{\substack{\alpha_1 \in \mathbb{N}_0 \\ \alpha_1 \leq l}} 2^{-\alpha_1 m}2^{-(l-\alpha_1)(m-1)} = 2^{-l(m-1)} \sum_{\substack{\alpha_1 \in \mathbb{N}_0 \\ \alpha_1 \leq l}} 2^{-\alpha_1}$$

$$\leq 2^{-l(m-1)}2,$$

where in the last step we just estimated geometric series. Now we assume that (6.3.4) holds for $d-1$, with $\alpha' := (\alpha_2, \ldots, \alpha_d)$ we obtain

$$\sum_{\substack{\alpha \in \mathbb{N}_0^d \\ |\alpha|=l}} 2^{-\sum_{k=1}^d \alpha_k(m-\delta_{k,d})} = \sum_{\substack{\alpha_1 \in \mathbb{N}_0 \\ \alpha_1 \leq l}} 2^{-\alpha_1 m} \sum_{\substack{\alpha' \in \mathbb{N}_0^{d-1} \\ |\alpha'|=l-\alpha_1}} 2^{-\sum_{k=2}^d \alpha_k(m-\delta_{k,d})}$$

$$= \sum_{\substack{\alpha_1 \in \mathbb{N}_0 \\ \alpha_1 \leq l}} 2^{-\alpha_1 m} \sum_{\substack{\alpha' \in \mathbb{N}_0^{d-1} \\ |\alpha'|=l-\alpha_1}} 2^{-\sum_{k=1}^{d-1} \alpha'_k(m-\delta_{k,d-1})}$$

$$\leq 2^{d-1} \sum_{\substack{\alpha_1 \in \mathbb{N}_0 \\ \alpha_1 \leq l}} 2^{-\alpha_1 m}2^{-(l-\alpha_1)(m-1)}$$

$$\leq 2^{d-1}2^{-l(m-1)} \sum_{\substack{\alpha_1 \in \mathbb{N}_0 \\ \alpha_1 \leq l}} 2^{-\alpha_1}$$

$$\lesssim 2^{-l(m-1)},$$

with the constant $C = 2^d$ in the last term. $\square$

Remark 6.12. Assuming uniformly geometric growth of the series, one gets a weaker decay with an additional polynomial factor:

$$\sum_{\substack{\alpha \in \mathbb{N}_0^d \\ |\alpha|=l}} 2^{-|\alpha|(m-1)} \lesssim 2^{-l(m-1)}(1+l)^{d-1}.$$

In the sequel we formulate auxiliary compression results for tensor product quarklets. Let the index set $\boldsymbol{\nabla}_\sigma$ be defined as in (5.1.10). With $\nabla_{j_i,\bar{\sigma}_i}$ as in (4.3.2) we analogously define the set of translation parameters

$$\boldsymbol{\nabla}_{\boldsymbol{j},\sigma} := \prod_{i=1}^d \nabla_{j_i,\bar{\sigma}_i}, \tag{6.3.5}$$

Proposition 6.13. *Let $m \geq 3$, $d \geq 2$. Let the weighted quarklets $(w_\lambda^{H^1})^{-1}\boldsymbol{\psi}_\lambda^\sigma$, $(w_{\lambda'}^{H^1})^{-1}\boldsymbol{\psi}_{\lambda'}^\sigma$, $\lambda := (\boldsymbol{p},\boldsymbol{j},\boldsymbol{k}) \in \boldsymbol{\nabla}_\sigma$, $\lambda' := (\boldsymbol{p}',\boldsymbol{j}',\boldsymbol{k}') \in \boldsymbol{\nabla}_\sigma$ be defined as in (5.1.11), and the bilinear form $a(\cdot,\cdot)$ as in (6.3.1). Then it holds*

$$|a((w_\lambda^{H^1})^{-1}\boldsymbol{\psi}_\lambda^\sigma, (w_{\lambda'}^{H^1})^{-1}\boldsymbol{\psi}_{\lambda'}^\sigma)| \lesssim \sum_{i=1}^d \prod_{r=1}^d \left(1+|p_r-p_r'|\right)^{m-1-\delta_2/2} 2^{-|j_r-j_r'|(m-1/2-\delta_{ir})}, \tag{6.3.6}$$

with δ_2 as in (5.1.14).

Proof. There is nothing to prove if $\operatorname{supp}\boldsymbol{\psi}_\lambda^\sigma \cap \operatorname{supp}\boldsymbol{\psi}_{\lambda'}^\sigma = \emptyset$. Otherwise we use the tensor product structure of the quarklets to obtain

$$a(\boldsymbol{\psi}_\lambda^\sigma,\boldsymbol{\psi}_{\lambda'}^\sigma) = \sum_{i=1}^d \prod_{r=1}^d \left\langle \left(\psi_{p_r,j_r,k_r}^{\sigma_r}\right)^{(\delta_{ir})}, \left(\psi_{p_r',j_r',k_r'}^{\sigma_r}\right)^{(\delta_{ir})} \right\rangle_{L_2(I)},$$

where the Kronecker deltas indicate whether the quarklet itself or its first derivative is concerned. Applying the estimates (6.2.4) and (6.2.12) leads to

$$|a(\boldsymbol{\psi}_\lambda^\sigma,\boldsymbol{\psi}_{\lambda'}^\sigma)| \leq \sum_{i=1}^d \prod_{r=1}^d \left((p_r+1)(p_r'+1)\right)^{m-1} 2^{\delta_{ir}(j_r+j_r')} 2^{-|j_r-j_r'|(m-1/2-\delta_{ir})}$$

$$\leq \sum_{i=1}^d 2^{j_i+j_i'} \prod_{r=1}^d \left((p_r+1)(p_r'+1)\right)^{m-1} 2^{-|j_r-j_r'|(m-1/2-\delta_{ir})}.$$

Estimating the weights $w_\lambda, w_{\lambda'}$ defined in (5.1.14) by the Cauchy-Schwarz inequality, we obtain

$$w_\lambda^{-1} w_{\lambda'}^{-1} \leq \left(\sum_{i=1}^d ((p_i+1)(p_i'+1))^{2+\delta_1/2} 2^{(j_i+j_i')}\right)^{-1} \prod_{r=1}^d ((p_r+1)(p_r'+1))^{-\delta_2/2}.$$

Combining the previous estimates, we obtain

$$|a(w_\lambda^{-1}\boldsymbol{\psi}_\lambda^\sigma, w_{\lambda'}^{-1}\boldsymbol{\psi}_{\lambda'}^\sigma)| \leq \sum_{i=1}^d \prod_{r=1}^d ((p_r+1)(p_r'+1))^{m-1-\delta_2/2} 2^{-|j_r-j_r'|(m-1/2-\delta_{ir})}.$$

Choosing $\delta_2 > 2m - 2$ and using the relation

$$(p + 1)(p' + 1) \geq 1 + |p - p'|,$$

we finally get the claim. □

In preparation for the compression estimates we make another auxiliary statement.

Proposition 6.14. *Let* $\boldsymbol{\lambda} := (\boldsymbol{p}, \boldsymbol{j}, \boldsymbol{k}) \in \boldsymbol{\nabla}_\sigma$ *and* $\boldsymbol{p}'$ *be fixed and let* $l \in \mathbb{N}$. *Let further*

$$a_{\boldsymbol{\lambda}, \boldsymbol{\lambda}'} := a((w_{\boldsymbol{\lambda}}^{H^1})^{-1} \boldsymbol{\psi}_{\boldsymbol{\lambda}}^\sigma, (w_{\boldsymbol{\lambda}'}^{H^1})^{-1} \boldsymbol{\psi}_{\boldsymbol{\lambda}'}^\sigma). \tag{6.3.7}$$

Then it holds

$$\sum_{\boldsymbol{j}' : |\boldsymbol{j} - \boldsymbol{j}'| = l} \sum_{\boldsymbol{k}' \in \nabla_{\boldsymbol{j}', \sigma}} 2^{|\boldsymbol{j}|/2} |a_{\boldsymbol{\lambda}, \boldsymbol{\lambda}'}| 2^{-|\boldsymbol{j}'|/2} \lesssim 2^{-l(m-2)} \prod_{r=1}^{d} \left(1 + |p_r - p_r'| \right)^{m-1-\delta_2/2}. \tag{6.3.8}$$

Proof. The number of indices $\boldsymbol{\lambda}' \in \boldsymbol{\nabla}_\sigma$ with fixed $\boldsymbol{p}'$ that fulfil $\mathrm{supp}\boldsymbol{\psi}_{\boldsymbol{\lambda}}^\sigma \cap \mathrm{supp}\boldsymbol{\psi}_{\boldsymbol{\lambda}'}^\sigma \neq \emptyset$ is of order $\prod_{i=1}^d \max\{1, 2^{j_i' - j_i}\}$. Furthermore, it holds

$$\prod_{i=1}^{d} \max\{1, 2^{j_i' - j_i}\} (2^{-|\boldsymbol{j}|/2})^{-1} 2^{-|\boldsymbol{j}'|/2} = 2^{|\boldsymbol{j} - \boldsymbol{j}'|/2}.$$

Combining this with (6.3.6) gives

$$\sum_{\boldsymbol{j}' : |\boldsymbol{j} - \boldsymbol{j}'| = l} \sum_{\boldsymbol{k}' \in \nabla_{\boldsymbol{j}', \sigma}} 2^{|\boldsymbol{j}|/2} |a_{\boldsymbol{\lambda}, \boldsymbol{\lambda}'}| 2^{-|\boldsymbol{j}'|/2}$$

$$\lesssim \sum_{\boldsymbol{j}' : |\boldsymbol{j} - \boldsymbol{j}'| = l} 2^{|\boldsymbol{j} - \boldsymbol{j}'|/2} |a_{\boldsymbol{\lambda}, \boldsymbol{\lambda}'}|$$

$$\lesssim \sum_{\boldsymbol{j}' : |\boldsymbol{j} - \boldsymbol{j}'| = l} \sum_{i=1}^{d} \prod_{r=1}^{d} \left(1 + |p_r - p_r'| \right)^{m-1-\delta_2/2} 2^{-|j_r - j_r'|(m-1-\delta_{ir})}$$

$$= \sum_{\boldsymbol{j}' : |\boldsymbol{j} - \boldsymbol{j}'| = l} \sum_{i=1}^{d} 2^{-\sum_{r=1}^{d} |j_r - j_r'|(m-1-\delta_{ir})} \left(\prod_{r=1}^{d} \left(1 + |p_r - p_r'| \right)^{m-1-\delta_2/2} \right)$$

$$= \sum_{i=1}^{d} \left(\sum_{\boldsymbol{j}' : |\boldsymbol{j} - \boldsymbol{j}'| = l} 2^{-\sum_{r=1}^{d} |j_r - j_r'|(m-1-\delta_{ir})} \right) \left(\prod_{r=1}^{d} \left(1 + |p_r - p_r'| \right)^{m-1-\delta_2/2} \right),$$

where we changed the order of summation for the last two equalities. Bounding the middle term by (6.3.4) we obtain

$$\sum_{\boldsymbol{j}' : |\boldsymbol{j} - \boldsymbol{j}'| = l} \sum_{\boldsymbol{k}'} 2^{|\boldsymbol{j}|/2} |a_{\boldsymbol{\lambda}, \boldsymbol{\lambda}'}| 2^{-|\boldsymbol{j}'|/2} \lesssim \sum_{i=1}^{d} 2^{-l(m-2)} \left(\prod_{r=1}^{d} \left(1 + |p_r - p_r'| \right)^{m-1-\delta_2/2} \right)$$

$$= 2^{-l(m-2)} \prod_{r=1}^{d} \left(1 + |p_r - p_r'| \right)^{m-1-\delta_2/2}.$$

□

6.3.1 First Compression

After all these preparations, we are able to state the first main result of this section.

Theorem 6.15. *Let $m \geq 3$. Let $\mathbf{A}$ the stiffness matrix of the Poisson equation (6.3.1) discretised by $\mathbf{\Psi}_\sigma^1$, defined in (5.1.13). Furthermore, for $J \in \mathbb{N}_0$, with $\lambda = (\boldsymbol{p}, \boldsymbol{j}, \boldsymbol{k})$, $\lambda' = (\boldsymbol{p}', \boldsymbol{j}', \boldsymbol{k}') \in \boldsymbol{\nabla}_\sigma$, define $\mathbf{A}_J$ by setting all entries from $\mathbf{A}$ to zero that satisfy*

$$a \log_2(\prod_{i=1}^{d} 1 + |p_i - p_i'|) + b|\boldsymbol{j} - \boldsymbol{j}'| > J, \tag{6.3.9}$$

where $a, b > 0$. Then, for $\delta_2 > 2m$, the maximal number of non-zero entries in each row and column of $\mathbf{A}_J$ is of the order

$$\left(J^{2d-2} 2^{\frac{J}{a}} + J^{d-1} 2^{\frac{J}{b}}\right) \begin{cases} J, & a = b, \\ 1, & otherwise. \end{cases} \tag{6.3.10}$$

Furthermore, with $\tau := m - 1 - \frac{\delta_2}{2}$ it holds that

$$\|\mathbf{A} - \mathbf{A}_J\|_{\mathcal{L}(\ell_2(\boldsymbol{\nabla}_\sigma))} \lesssim \left(2^{-(m-2)\frac{J}{b}} + J^{d-1} 2^{(1+\tau)\frac{J}{a}}\right) \begin{cases} J, & \frac{a}{b} = -\frac{1+\tau}{m-2}, \\ 1, & otherwise. \end{cases} \tag{6.3.11}$$

The proof of Theorem 6.15 is quite technical. In the course of the proof, we will use the following facts:

(i) Let $K \in \mathbb{N}$, $t \in \mathbb{R}_+$. Then,

$$\sum_{n=1}^{K} n^{-t} \leq 1 + \int_1^K x^{-t} \mathrm{d}x \lesssim \begin{cases} K^{1-t}, & t < 1, \\ 1 + \ln(K), & t = 1, \\ 1, & t > 1. \end{cases} \tag{6.3.12}$$

(ii) Let $K \in \mathbb{N}$, $t > 1$. Then,

$$\sum_{n=K}^{\infty} n^{-t} \leq K^{-t} + \int_K^\infty x^{-t} \mathrm{d}x \lesssim K^{1-t}. \tag{6.3.13}$$

Proof of Theorem 6.15. First we are going to estimate the number of non-trivial entries, i.e., (6.3.10). To simplify the notation we assume $j_0 = 0$ for the minimal level in each coordinate of the quarklet frame $\mathbf{\Psi}_\sigma^1$.

Let $\lambda \in \boldsymbol{\nabla}_\sigma$ be fixed. The number of $\lambda' \in \boldsymbol{\nabla}_\sigma$ with fixed $\boldsymbol{p}'$ that fulfil $\mathrm{supp}\psi_\lambda^\varrho \cap \mathrm{supp}\psi_{\lambda'}^\sigma \neq \emptyset$ is of the order $\prod_{i=1}^{d} \max\{1, 2^{j_i' - j_i}\} \leq 2^{|\boldsymbol{j} - \boldsymbol{j}'|}$. Further, $|\{\boldsymbol{j} \in \mathbb{N}_0^d : |\boldsymbol{j}| =$

$l\}| = \binom{l+d-1}{l} \lesssim (1+l)^{d-1}$ with a constant depending on d holds. Together, this implies that the number of entries in the λ-th row of $\mathbf{A}_J$ is bounded by

$$
\sum_{\substack{\boldsymbol{p'} \in \mathbb{N}_0^d \\ \prod_{i=1}^{d} 1+|p_i-p_i'| \leq 2^{\frac{J}{a}}}} \sum_{l=0}^{\lfloor \frac{J}{b} - \frac{a}{b} \log_2(\prod_{i=1}^{d} 1+|p_i-p_i'|) \rfloor} \sum_{\substack{\boldsymbol{j'} \in \mathbb{N}_0^d \\ |j-j'|=l}} 2^{|j-j'|}
$$

$$
\leq \sum_{\substack{\boldsymbol{p''} \in \mathbb{N}^d \\ \prod_{i=1}^{d} p_i'' \leq 2^{\frac{J}{a}}}} \sum_{l=0}^{\lfloor \frac{J}{b} - \frac{a}{b} \log_2(\prod_{i=1}^{d} p_i'') \rfloor} \binom{l+d-1}{l} 2^l.
$$

In the latter term, $\binom{l+d-1}{l}$ can be estimated from above by $\left(1 + \frac{J}{b}\right)^{d-1}$. Hence,

$$
\sum_{\substack{\boldsymbol{p'} \in \mathbb{N}_0^d \\ \prod_{i=1}^{d} 1+|p_i-p_i'| \leq 2^{\frac{J}{a}}}} \sum_{l=0}^{\lfloor \frac{J}{b} - \frac{a}{b} \log_2(\prod_{i=1}^{d} 1+|p_i-p_i'|) \rfloor} \sum_{\substack{\boldsymbol{j'} \in \mathbb{N}_0^d \\ |j-j'|=l}} 2^{|j-j'|}
$$

$$
\lesssim \left(\frac{J}{b}\right)^{d-1} 2^{\frac{J}{b}} \sum_{\substack{\boldsymbol{p''} \in \mathbb{N}^d \\ \prod_{i=1}^{d} p_i'' \leq 2^{\frac{J}{a}}}} \left(\prod_{i=1}^{d} p_i''\right)^{-\frac{a}{b}}.
$$

(6.3.14)

We separate the last component of $\boldsymbol{p''}$ to obtain

$$
\sum_{\substack{\boldsymbol{p''} \in \mathbb{N}^d \\ \prod_{i=1}^{d} p_i'' \leq 2^{\frac{J}{a}}}} \left(\prod_{i=1}^{d} p_i''\right)^{-\frac{a}{b}} = \sum_{\substack{\boldsymbol{p''} \in \mathbb{N}^{d-1} \\ \prod_{i=1}^{d-1} p_i'' \leq 2^{\frac{J}{a}}}} \sum_{p_d''=1}^{2^{\frac{J}{a}} \left(\prod_{i=1}^{d-1} p_i''\right)^{-1}} \left(\prod_{i=1}^{d} p_i''\right)^{-\frac{a}{b}}.
$$

Applying (6.3.12) d times with $K = 2^{J/a}$ leads to

$$
\sum_{\substack{\boldsymbol{p''} \in \mathbb{N}^d \\ \prod_{i=1}^{d} p_i'' \leq 2^{\frac{J}{a}}}} \left(\prod_{i=1}^{d} p_i''\right)^{-\frac{a}{b}} \lesssim \sum_{\substack{\boldsymbol{p''} \in \mathbb{N}^{d-1} \\ \prod_{i=1}^{d-1} p_i'' \leq 2^{\frac{J}{a}}}} \begin{cases} 2^{\frac{J}{a}(1-\frac{a}{b})} \left(\prod_{i=1}^{d-1} p_i''\right)^{-1}, & a < b, \\ \left(1 + \frac{J}{a} - \ln(\prod_{i=1}^{d-1} p_i'')\right) \left(\prod_{i=1}^{d-1} p_i''\right)^{-1}, & a = b, \\ \left(\prod_{i=1}^{d-1} p_i''\right)^{-1}, & a > b, \end{cases}
$$

$$
\lesssim \begin{cases} 2^{\frac{J}{a}(1-\frac{a}{b})} \left(1 + \frac{J}{a}\right)^{d-1}, & a < b, \\ \left(1 + \frac{J}{a}\right)^{d}, & a = b, \\ 1, & a > b. \end{cases}
$$

(6.3.15)

Finally, by the last estimate, (6.3.14) can be further estimated by

$$\sum_{\substack{p'' \in \mathbb{N}^d \\ \prod_{i=1}^d p_i'' \leq 2^{\frac{J}{a}}}} \left(\frac{J}{b}\right)^{d-1} 2^{\frac{J}{b}} \left(\prod_{i=1}^d p_i''\right)^{-\frac{a}{b}} \lesssim \begin{cases} \left(\frac{J}{b}\right)^{d-1} 2^{\frac{J}{a}} \left(1 + \frac{J}{a}\right)^{d-1}, & a < b, \\ \left(\frac{J}{b}\right)^{d-1} 2^{\frac{J}{b}} \left(1 + \frac{J}{a}\right)^d, & a = b, \\ \left(\frac{J}{b}\right)^{d-1} 2^{\frac{J}{b}}, & a > b, \end{cases}$$

which implies (6.3.10). Next we will derive the compression result (6.3.11). Again, we use Lemma 6.2. Analogously as in Section 6.2 it is sufficient to show

$$\sup_{\lambda \in \nabla_\sigma} \alpha_\lambda \lesssim \left(2^{-(m-2)\frac{J}{b}} + J^{d-1} 2^{(1+\tau)\frac{J}{a}}\right) \begin{cases} J, & \frac{a}{b} = -\frac{1+\tau}{m-2}, \\ 1, & \text{otherwise}, \end{cases}$$

where

$$\alpha_\lambda := w_\lambda^{-1} \sum_{\lambda' \in \nabla_\sigma} |(\mathbf{A})_{\lambda,\lambda'} - (\mathbf{A}_J)_{\lambda,\lambda'}| w_{\lambda'}. \tag{6.3.16}$$

Separating the level and polynomial degree in (6.3.9) and with $\beta := \lceil b^{-1}(J - a \log_2(\prod_{i=1}^d 1 + |p_i - p_i'|))\rceil$ we obtain

$$\alpha_\lambda \lesssim \sum_{p' \in \mathbb{N}_0^d} \sum_{l=\max\{0,\beta\}}^\infty \sum_{\substack{j' \in \mathbb{N}_0^d \\ |j-j'|=l}} \sum_{k' \in \nabla_{j',\sigma}} 2^{|j|/2} |a_{\lambda,\lambda'}| 2^{-|j'|/2},$$

where we chose weights of the form $w_\lambda = 2^{-|j|/2}$. With (6.3.8) we get

$$\alpha_\lambda \lesssim \sum_{p' \in \mathbb{N}_0^d} \left(\prod_{r=1}^d \left(1 + |p_r - p_r'|\right)^{m-1-\delta_2/2}\right) \sum_{l=\max\{0,\beta\}}^\infty 2^{-l(m-2)}.$$

Splitting the sum yields

$$\alpha_\lambda \lesssim \sum_{\substack{p' \in \mathbb{N}_0^d \\ \beta \leq 0}} \prod_{i=1}^d \left(1 + |p_i - p_i'|\right)^\tau$$

$$+ \sum_{\substack{p' \in \mathbb{N}_0^d \\ \beta > 0}} \prod_{i=1}^d \left(1 + |p_i - p_i'|\right)^\tau 2^{-(m-2)\beta}. \tag{6.3.17}$$

First we have a closer look at the first sum of (6.3.17). By splitting the sum we get

$$\sum_{\substack{p' \in \mathbb{N}_0^d \\ \beta \leq 0}} \prod_{i=1}^d \left(1 + |p_i - p_i'|\right)^\tau = \sum_{\substack{p' \in \mathbb{N}_0^d \\ \log_2(\prod_{i=1}^d 1 + |p_i - p_i'|) \geq J/a}} \prod_{i=1}^d (1 + |p_i - p_i'|)^\tau$$

$$= \sum_{p' \in \mathbb{N}_0^{d-1}} \prod_{i=1}^{d-1} (1 + |p_i - p_i'|)^\tau \sum_{\substack{p_d' \in \mathbb{N}_0 \\ \log_2(1+|p_d-p_d'|) \geq J/a - \log_2(\prod_{i=1}^{d-1} 1 + |p_i - p_i'|)}} (1 + |p_d - p_d'|)^\tau.$$

Again splitting the sum leads to

$$\sum_{\substack{p' \in \mathbb{N}_0^d \\ \beta \leq 0}} \prod_{i=1}^{d} (1 + |p_i - p_i'|)^\tau$$

$$\lesssim \sum_{\substack{p' \in \mathbb{N}_0^{d-1} \\ \log_2(\prod_{i=1}^{d-1} 1 + |p_i - p_i'|) \geq J/a}} \prod_{i=1}^{d-1} (1 + |p_i - p_i'|)^\tau \sum_{\substack{p_d' \in \mathbb{N}_0 \\ \log_2(1 + |p_d - p_d'|) \geq 0}} (1 + |p_d - p_d'|)^\tau$$

$$+ \sum_{\substack{p' \in \mathbb{N}_0^{d-1} \\ \log_2(\prod_{i=1}^{d-1} 1 + |p_i - p_i'|) < J/a}} \prod_{i=1}^{d-1} (1 + |p_i - p_i'|)^\tau$$

$$\sum_{\substack{p_d' \in \mathbb{N}_0 \\ \log_2(1 + |p_d - p_d'|) \geq J/a - \log_2(\prod_{i=1}^{d-1} 1 + |p_i - p_i'|)}} (1 + |p_d - p_d'|)^\tau.$$

Consequently, with (6.3.13) with $t = -\tau$, $K = 1$ and $K = 2^{J/a - \log_2(\prod_{i=1}^{d} 1 + |p_i - p_i'|)}$, respectively, we get

$$\sum_{\substack{p' \in \mathbb{N}_0^d \\ \beta \leq 0}} \prod_{i=1}^{d} (1 + |p_i - p_i'|)^\tau$$

$$\lesssim \sum_{\substack{p' \in \mathbb{N}_0^{d-1} \\ \log_2(\prod_{i=1}^{d-1} 1 + |p_i - p_i'|) \geq J/a}} \prod_{i=1}^{d-1} (1 + |p_i - p_i'|)^\tau 1^{1+\tau}$$

$$+ \sum_{\substack{p' \in \mathbb{N}_0^{d-1} \\ \log_2(\prod_{i=1}^{d-1} 1 + |p_i - p_i'|) < J/a}} \prod_{i=1}^{d-1} (1 + |p_i - p_i'|)^\tau 2^{(1+\tau)(J/a - \log_2(\prod_{i=1}^{d} 1 + |p_i - p_i'|))}.$$

Putting all together gives

$$\sum_{\substack{p' \in \mathbb{N}_0^d \\ \beta \leq 0}} \prod_{i=1}^{d} (1 + |p_i - p_i'|)^\tau = \sum_{\substack{p' \in \mathbb{N}_0^d \\ \log_2(\prod_{i=1}^{d} 1 + |p_i - p_i'|) \geq J/a}} \prod_{i=1}^{d} (1 + |p_i - p_i'|)^\tau$$

$$\lesssim \sum_{\substack{p' \in \mathbb{N}_0^{d-1} \\ \log_2(\prod_{i=1}^{d-1} 1 + |p_i - p_i'|) \geq J/a}} \prod_{i=1}^{d-1} (1 + |p_i - p_i'|)^\tau$$

$$+ \sum_{\substack{p' \in \mathbb{N}_0^{d-1} \\ \log_2(\prod_{i=1}^{d-1} 1 + |p_i - p_i'|) < J/a}} \prod_{i=1}^{d-1} (1 + |p_i - p_i'|)^{-1} 2^{(1+\tau)J/a}.$$

It follows by induction and with an estimate similar as in (6.3.15), that

$$\sum_{\substack{p' \in \mathbb{N}_0^d \\ \beta \leq 0}} \prod_{i=1}^{d} (1 + |p_i - p_i'|)^{\tau} \lesssim 2^{(1+\tau)J/a} (1 + \frac{J}{a})^{d-1}. \tag{6.3.18}$$

Now we turn to the second sum of (6.3.17). With the definition of β we obtain

$$\sum_{\substack{p' \in \mathbb{N}_0^d \\ \beta > 0}} \prod_{i=1}^{d} \left(1 + |p_i - p_i'|\right)^{\tau} 2^{-(m-2)\beta}$$

$$= \sum_{\substack{p' \in \mathbb{N}_0^d \\ \log_2(\prod_{i=1}^{d} 1 + |p_i - p_i'|) < J/a}} \prod_{i=1}^{d} \left(1 + |p_i - p_i'|\right)^{\tau + (m-2)\frac{a}{b}} 2^{-(m-2)J/b}$$

Similar estimates as in (6.3.15) imply

$$\sum_{\substack{p' \in \mathbb{N}_0^d \\ \beta > 0}} \prod_{i=1}^{d} \left(1 + |p_i - p_i'|\right)^{\tau} 2^{-(m-2)\beta}$$

$$\lesssim 2^{-(m-2)\frac{J}{b}} \begin{cases} 2^{(1+\tau+(m-2)\frac{a}{b})J/a} (1 + \frac{J}{a})^{d-1}, & \tau + (m-2)\frac{a}{b} > -1, \\ (1 + \frac{J}{a})^d, & \tau + (m-2)\frac{a}{b} = -1, \\ 1, & \tau + (m-2)\frac{a}{b} < -1, \end{cases}$$

$$\lesssim \left(2^{-(m-2)\frac{J}{b}} + \left(1 + \frac{J}{a}\right)^{d-1} 2^{(1+\tau)\frac{J}{a}}\right)$$

$$\cdot \begin{cases} (1 + \frac{J}{a}), & \tau + (m-2)\frac{a}{b} = -1, \\ 1, & \text{otherwise.} \end{cases} \tag{6.3.19}$$

Finally, combining (6.3.17) - (6.3.19) yields (6.3.11). □

6.3.2 Second Compression

In the following we discuss second compression ideas in the multivariate case. We have additional freedom in the design of cut-off rules since the singular supports may not intersect for all coordinate directions. In particular, we distinguish between the following three cases. If all singular supports are intersected, we never apply second compression. If no singular supports intersect, we apply second compression. If $1, \ldots, d-1$ singular supports are intersected, we consider two versions of compression. In Theorem 6.17 we consider a *weak* second compression strategy that drops just matrix entries where all singular supports are not intersected. In Theorem 6.18 we consider a *strong* second compression strategy that drops all matrix entries where one singular support is not intersected. We begin with an auxiliary statement.

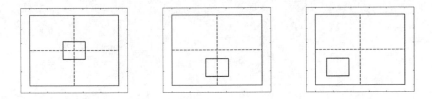

Figure 6.3: Cases of intersection of the supports. Lines indicate the singular support in one coordinate direction.

Proposition 6.16. *Let* $\boldsymbol{\lambda} := (\boldsymbol{p}, \boldsymbol{j}, \boldsymbol{k})$ *and* $\boldsymbol{\lambda}' := (\boldsymbol{p}', \boldsymbol{j}', \boldsymbol{k}')$ *so that*

$$\lambda'_r \in \bigcup_{j'_r, i} A_{j'_r, \lambda_r, i} \quad \forall r = 1, \ldots, d. \tag{6.3.20}$$

Then it holds

$$|a((w_{\boldsymbol{\lambda}}^{H^1})^{-1}\boldsymbol{\psi}_{\boldsymbol{\lambda}}^{\sigma}, (w_{\boldsymbol{\lambda}'}^{H^1})^{-1}\boldsymbol{\psi}_{\boldsymbol{\lambda}'}^{\sigma})| \lesssim \sum_{i=1}^{d} \prod_{r=1}^{d} \left(1 + |p_r - p'_r|\right)^{m+1-\delta_2/2} 2^{-(j'_r - j_r)(m+1/2-\delta_{ir})}, \tag{6.3.21}$$

Furthermore, for fixed $\boldsymbol{\lambda} := (\boldsymbol{p}, \boldsymbol{j}, \boldsymbol{k})$ *it holds*

$$\sum_{\substack{\boldsymbol{j}':|\boldsymbol{j}-\boldsymbol{j}'|=l \\ |j_r - j'_r| > \log(m)+1 \\ \forall r = 1,\ldots,d}} \sum_{\substack{k' \in \boldsymbol{\nabla}_{j',\sigma} \\ \lambda'_r \in \bigcup_{j'_r, i} A_{j'_r, \lambda_r, i} \\ \forall r = 1,\ldots,d}} 2^{|\boldsymbol{j}|/2}|a_{\boldsymbol{\lambda}, \boldsymbol{\lambda}'}|2^{-|\boldsymbol{j}'|/2} \lesssim 2^{-l(m-1)} \prod_{r=1}^{d} \left(1 + |p_r - p'_r|\right)^{m+1-\delta_2/2}. \tag{6.3.22}$$

Proof. The proof is completely analogous to the proof of (6.3.6) and (6.3.8), but instead of (6.2.4), (6.2.12) we use (6.3.2), (6.3.3). In particular the asymptotic number of quarklets is the same. $\qquad\square$

As aforementioned, the following theorem states the effect of weak second compression and is another main result of this section.

Theorem 6.17. *Let* $m \geq 3$. *Let* $\mathbf{A}$ *the stiffness matrix of the Poisson equation* (6.3.1) *discretised by* $\boldsymbol{\Psi}_{\sigma}^1$, *defined in* (5.1.13). *Further, for* $J_1, J_2 \in \mathbb{N}_0, J_1 > J_2$, *with* $\boldsymbol{\lambda} = (\boldsymbol{p}, \boldsymbol{j}, \boldsymbol{k}), \boldsymbol{\lambda}' = (\boldsymbol{p}', \boldsymbol{j}', \boldsymbol{k}') \in \boldsymbol{\nabla}_{\sigma}$, *define* $\mathbf{A}_J$ *by setting all entries from* $\mathbf{A}$ *to zero that satisfy* (6.3.9) *with respect to* J_1 *or*

$$J_1 \geq a \log_2(\prod_{i=1}^{d} 1 + |p_i - p'_i|) + b|\boldsymbol{j} - \boldsymbol{j}'| > J_2,$$

$$\lambda'_r \in \bigcup_{j'_r, i} A_{j'_r, \lambda_r, i} \quad \vee \quad \lambda_r \in \bigcup_{j_r, i} A_{j_r, \lambda'_r, i} \quad \forall r = 1, \ldots, d. \tag{6.3.23}$$

where $a, b > 0$. Then , for $\delta_2 > 2m + 4$, the maximal number of non-zero entries in each row and column of $\mathbf{A}_J$ is of the order

$$
\left(J_2^{2d-2} 2^{\frac{J_2}{a}} + J_2^{d-1} 2^{\frac{J_2}{b}} \right) \begin{cases} J_2, & a = b, \\ 1, & \text{otherwise.} \end{cases}
$$
$$
+ \left(J_1^{2d-4} 2^{\frac{J_1}{a}} + J_1^{d-1} 2^{\frac{J_1}{b}} \right) \begin{cases} J_1, & a = b, \\ 1, & \text{otherwise.} \end{cases}
\tag{6.3.24}
$$

Furthermore, with $\tau := m + 1 - \frac{\delta_2}{2}$ it holds that

$$
\|\mathbf{A} - \mathbf{A}_J\|_{\mathcal{L}(\ell_2(\boldsymbol{\nabla}_\sigma))} \lesssim \left(2^{-(m-2)\frac{J_1}{b}} + J_1^{d-1} 2^{(-1+\tau)\frac{J_1}{a}} \right) \begin{cases} J_1, & \frac{a}{b} = -\frac{-1+\tau}{m-2}, \\ 1, & \text{otherwise.} \end{cases}
$$
$$
+ \left(2^{-(m-1)\frac{J_2}{b}} + J_1^{d-1} 2^{(1+\tau)\frac{J_1}{a}} \right) \begin{cases} J_1, & \frac{a}{b} = -\frac{1+\tau}{m-1}, \\ 1, & \text{otherwise.} \end{cases}
\tag{6.3.25}
$$

Proof. The proof is similar to the proofs of Theorem 6.15 and 6.7. First we consider the number of entries. Analogous to the univariate setting, the case $a \log_2(\prod_{i=1}^d 1 + |p_i - p_i'|) + b|\boldsymbol{j} - \boldsymbol{j}'| \leq J_2$ has already been considered, cf. (6.3.10). Hence, for this case the asymptotic number of entries is bounded by

$$
\left(J_2^{2d-2} 2^{\frac{J_2}{a}} + J_2^{d-1} 2^{\frac{J_2}{b}} \right) \begin{cases} J_2, & a = b, \\ 1, & \text{otherwise.} \end{cases}
$$

Hence we just have to consider the case $J_2 < a \log_2(\prod_{i=1}^d 1 + |p_i - p_i'|) + b|\boldsymbol{j} - \boldsymbol{j}'| \leq J_1$ for quarklets with intersecting singular support or small scale difference. Let $\boldsymbol{\lambda}'$ fixed. Then the number of non-zero-entries is given by

$$
N := \sum_{\substack{J_2 < a \log_2(\prod_{i=1}^d 1 + |p_i - p_i'|) + b|\boldsymbol{j} - \boldsymbol{j}'| \leq J_1 \\ \exists r \in \{1,\dots,d\}: \lambda_r \notin \bigcup_{j_r,i} A_{j_r, \lambda'_r, i}}} 1
$$
$$
\lesssim \sum_{\substack{\boldsymbol{p}' \in \mathbb{N}_0^d \\ \prod_{i=1}^d 1 + |p_i - p_i'| \leq 2^{\frac{J_1}{a}}}} \sum_{l=\beta_2}^{\beta_1 - 1} \sum_{\substack{\boldsymbol{j}' \in \mathbb{N}_0^d \\ |\boldsymbol{j} - \boldsymbol{j}'| = l}} \sum_{\substack{k \\ \exists r \in \{1,\dots,d\}: \lambda_r \notin \bigcup_{j_r,i} A_{j_r, \lambda'_r, i}}} 1.
$$

We additionally have to distinguish between coarse and far levels $\boldsymbol{j}$. In our case, for $|j_r - j_r'| < \log_2(m) + 1$ second compression ideas can never be applied. On coarse levels the number of intersecting quarklets is of order $2^{|j - j'|}$, on higher levels the number of quarklets intersecting the singular support of $\psi_{\lambda'}$ is of order $2^{|j - j'| - |j_r - j_r'|}$.

Accordingly splitting the sum leads to

$$
N \lesssim \sum_{\substack{p' \in \mathbb{N}_0^d \\ \prod_{i=1}^{d} 1 + |p_i - p_i'| \leq 2^{\frac{J_1}{a}}}} \sum_{l=\beta_2}^{\beta_1 - 1} \sum_{\substack{j' \in \mathbb{N}_0^d \\ |j - j'| = l}} 2^{|j - j'| - |j_r - j_r'|}
$$

$$
+ \sum_{\substack{p' \in \mathbb{N}_0^d \\ \prod_{i=1}^{d} 1 + |p_i - p_i'| \leq 2^{\frac{J_1}{a}}}} \sum_{l=0}^{d(\log_2(m)+1)} \sum_{\substack{j' \in \mathbb{N}_0^d \\ |j - j'| = l}} 2^{|j - j'|}.
$$

We have a closer look at the first sum:

$$
\sum_{\substack{p' \in \mathbb{N}_0^d \\ \prod_{i=1}^{d} 1 + |p_i - p_i'| \leq 2^{\frac{J_1}{a}}}} \sum_{l=\beta_2}^{\beta_1 - 1} \sum_{\substack{j' \in \mathbb{N}_0^d \\ |j - j'| = l}} 2^{|j - j'| - |j_r - j_r'|}
$$

$$
\lesssim \sum_{\substack{p' \in \mathbb{N}_0^d \\ \prod_{i=1}^{d} 1 + |p_i - p_i'| \leq 2^{\frac{J_1}{a}}}} \sum_{l=\beta_2}^{\beta_1 - 1} 2^l (1 + l)^{d-2}
$$

$$
= \sum_{\substack{p' \in \mathbb{N}_0^d \\ \prod_{i=1}^{d} 1 + |p_i - p_i'| \leq 2^{\frac{J_1}{a}}}} \sum_{l=0}^{\beta_1 - \beta_2 - 1} 2^{l+\beta_2} (1 + l + \beta_2)^{d-2}
$$

$$
\lesssim \sum_{\substack{p' \in \mathbb{N}_0^d \\ \prod_{i=1}^{d} 1 + |p_i - p_i'| \leq 2^{\frac{J_1}{a}}}} 2^{J_2/b} 2^{\frac{J_1 - J_2}{b}} (1 + \frac{J_1}{b})^{d-2} (\prod_{i=1}^{d} 1 + |p_i - p_i'|)^{-\frac{a}{b}}
$$

Similar estimates as in (6.3.15) lead to

$$
\sum_{\substack{p'' \in \mathbb{N}^d \\ \prod_{i=1}^{d} p_i'' \leq 2^{\frac{J_1}{a}}}} \left(\frac{J_1}{b}\right)^{d-2} 2^{\frac{J_2}{b}} 2^{\frac{J_1 - J_2}{b}} \left(\prod_{i=1}^{d} p_i''\right)^{-\frac{a}{b}} \lesssim
\begin{cases}
\left(\frac{J_1}{b}\right)^{d-2} 2^{\frac{J_1}{a}} \left(1 + \frac{J_1}{a}\right)^{d-1}, & a < b, \\
\left(\frac{J_1}{b}\right)^{d-2} 2^{\frac{J_1}{b}} \left(1 + \frac{J_1}{a}\right)^{d}, & a = b, \\
\left(\frac{J_1}{b}\right)^{d-2} 2^{\frac{J_1}{b}}, & a > b,
\end{cases}
$$

Similar computations for second sum give an upper bound of the form $2^{J_1/a}(1 + \frac{J_1}{a})^{d-1}$, where the inner sums are bounded by a constant. Now we turn to the compression result. The asymptotic error for $a \log_2(\prod_{i=1}^{d} 1 + |p_i - p_i'|) + b|j - j'| > J_1$ is given by (6.3.11). Again we use the Schur lemma, i.e., we consider

$$
\alpha_\lambda := w_\lambda^{-1} \sum_{\lambda' \in \nabla_\sigma} |(\mathbf{A})_{\lambda, \lambda'} - (\mathbf{A}_J)_{\lambda, \lambda'}| w_{\lambda'}.
$$

Separating the level and polynomial degree in (6.3.23) and with $\beta_n := \lceil b^{-1}(J_n - a \log_2(\prod_{i=1}^{d} 1 + |p_i - p_i'|))\rceil, n = 1, 2$ we obtain

$$
\alpha_\lambda \lesssim \sum_{\substack{p' \in \mathbb{N}_0^d \\ \prod_{i=1}^{d} 1 + |p_i - p_i'| \leq 2^{J_1/a}}} \sum_{l=\beta_2}^{\beta_1} \sum_{\substack{j':|j-j'|=l \\ |j_r - j_r'| > \log(m)+1 \\ \forall r=1,\ldots,d}} \sum_{\substack{k' \in \nabla_{j',\sigma} \\ \lambda_r' \in \bigcup_{j_r',i} A_{j_r',\lambda_r,i} \\ \forall r=1,\ldots,d}} 2^{|j|/2} |a_{\lambda,\lambda'}| 2^{-|j'|/2},
$$

where we chose weights of the form $w_\lambda = 2^{-|j|/2}$. With (6.3.22) we get

$$
\alpha_\lambda \lesssim \sum_{\substack{p' \in \mathbb{N}_0^d \\ \prod_{i=1}^{d} 1 + |p_i - p_i'| \leq 2^{J_1/a}}} \left(\prod_{r=1}^{d} \left(1 + |p_r - p_r'|\right) \right)^{m+1-\delta_2/2} \sum_{l=\beta_2}^{\beta_1} 2^{-l(m-1)}.
$$

Performing an index transformation we obtain

$$
\alpha_\lambda \lesssim \sum_{\substack{p' \in \mathbb{N}_0^d \\ \prod_{i=1}^{d} 1 + |p_i - p_i'| \leq 2^{J_1/a}}} \left(\prod_{r=1}^{d} \left(1 + |p_r - p_r'|\right) \right)^{m+1-\delta_2/2} \sum_{l=0}^{\beta_1 - \beta_2} 2^{-(l+\beta_2)(m-1)}
$$

$$
\lesssim \sum_{\substack{p' \in \mathbb{N}_0^d \\ \prod_{i=1}^{d} 1 + |p_i - p_i'| \leq 2^{J_1/a}}} \left(\prod_{r=1}^{d} \left(1 + |p_r - p_r'|\right) \right)^{\tau + (m-1)\frac{a}{b}} 2^{-(m-1)J_2/b}.
$$

Similar estimates as in (6.3.15) imply

$$
\alpha_\lambda \lesssim 2^{-(m-1)\frac{J_2}{b}} \begin{cases} 2^{(1+\tau+(m-1)\frac{a}{b})J_1/a}(1+\frac{J_1}{a})^{d-1}, & \tau + (m-1)\frac{a}{b} > -1, \\ (1 + \frac{J_1}{a})^d, & \tau + (m-1)\frac{a}{b} = -1, \\ 1, & \tau + (m-1)\frac{a}{b} < -1, \end{cases}
$$

$$
\lesssim \left(2^{-(m-1)\frac{J_2}{b}} + \left(1 + \frac{J_1}{a}\right)^{d-1} 2^{(1+\tau)\frac{J_1}{a}} \right)
$$

$$
\cdot \begin{cases} (1 + \frac{J_1}{a}), & \tau + (m-1)\frac{a}{b} = -1, \\ 1, & \text{otherwise.} \end{cases}
$$

$\square$

Finally, the following theorem states the effect of strong second compression and is another main result of this section.

Theorem 6.18. *Let $m \geq 3$. Let $\mathbf{A}$ the stiffness matrix of the Poisson equation (6.3.1) discretised by $\mathbf{\Psi}_\sigma^1$, defined in (5.1.13). Furthermore, for $J_1, J_2 \in \mathbb{N}_0, J_1 > J_2$, with $\boldsymbol{\lambda} = (\boldsymbol{p}, \boldsymbol{j}, \boldsymbol{k}), \boldsymbol{\lambda}' = (\boldsymbol{p}', \boldsymbol{j}', \boldsymbol{k}') \in \boldsymbol{\nabla}_\sigma$, define $\mathbf{A}_J$ by setting all entries from $\mathbf{A}$ to zero that satisfy (6.3.9) with respect to J_1 or*

$$J_1 \geq a \log_2(\prod_{i=1}^d 1 + |p_i - p'_i|) + b|\boldsymbol{j} - \boldsymbol{j}'| > J_2,$$

$$\exists r \in \{1, \ldots, d\} : \lambda'_r \in \bigcup_{j'_r, i} A_{j'_r, \lambda_r, i} \quad \vee \quad \lambda_r \in \bigcup_{j_r, i} A_{j_r, \lambda'_r, i}. \tag{6.3.26}$$

where $a, b > 0$. Then, for $\delta_2 > 2m$, the maximal number of non-zero entries in each row and column of $\mathbf{A}_J$ is of the order

$$\left(J_2^{2d-2} 2^{\frac{J_2}{a}} + J_2^{d-1} 2^{\frac{J_2}{b}} \right) \begin{cases} J_2, & a = b, \\ 1, & \textit{otherwise.} \end{cases}$$

$$+ J_1^{d-1} J_2^{d-1} 2^{\frac{J_1}{a}} \tag{6.3.27}$$

Furthermore, with $\tau := m - 1 - \frac{\delta_2}{2}$ it holds that

$$\|\mathbf{A} - \mathbf{A}_J\|_{\mathcal{L}(\ell_2(\boldsymbol{\nabla}_\sigma))} \lesssim \left(2^{-(m-2)\frac{J_1}{b}} + J_1^{d-1} 2^{(1+\tau)\frac{J_1}{a}} \right) \begin{cases} J_1, & \frac{a}{b} = -\frac{1+\tau}{m-2}, \\ 1, & \textit{otherwise.} \end{cases}$$

$$+ \left(2^{-(m-2)\frac{J_2}{b}} + J_1^{d-1} 2^{(1+\tau)\frac{J_1}{a}} \right) \begin{cases} J_1, & \frac{a}{b} = -\frac{1+\tau}{m-2}, \\ 1, & \textit{otherwise.} \end{cases} \tag{6.3.28}$$

Proof. First we consider the number of entries. Analogously to the univariate setting, the case $a \log_2(\prod_{i=1}^d 1 + |p_i - p'_i|) + b|\boldsymbol{j} - \boldsymbol{j}'| \leq J_2$ has already been considered, cf. (6.3.10). Hence, for this case the asymptotic number of entries is bounded by

$$\left(J_2^{2d-2} 2^{\frac{J_2}{a}} + J_2^{d-1} 2^{\frac{J_2}{b}} \right) \begin{cases} J_2, & a = b, \\ 1, & \text{otherwise.} \end{cases}$$

Thus, we just have to consider the case $J_2 < a \log_2(\prod_{i=1}^d 1 + |p_i - p'_i|) + b|\boldsymbol{j} - \boldsymbol{j}'| \leq J_1$ for quarklets with intersecting singular support or small scale difference. Let $\boldsymbol{\lambda}'$ fixed. Then the number of non-zero-entries is given by

$$N := \sum_{\substack{J_2 < a \log_2(\prod_{i=1}^d 1 + |p_i - p'_i|) + b|\boldsymbol{j} - \boldsymbol{j}'| \leq J_1 \\ \lambda_r \notin \bigcup_{j_r, i} A_{j_r, \lambda'_r, i}, \forall r = 1, \ldots, d}} 1$$

$$\lesssim \sum_{\substack{\boldsymbol{p}' \in \mathbb{N}_0^d \\ \prod_{i=1}^d 1 + |p_i - p'_i| \leq 2^{\frac{J_1}{a}}}} \sum_{l = \beta_2}^{\beta_1 - 1} \sum_{\substack{\boldsymbol{j}' \in \mathbb{N}_0^d \\ |\boldsymbol{j} - \boldsymbol{j}'| = l}} \sum_{\substack{k \\ \lambda_r \notin \bigcup_{j_r, i} A_{j_r, \lambda'_r, i}, \forall r = 1, \ldots, d}} 1.$$

We additionally have to distinguish between coarse and far levels $\boldsymbol{j}$. In our case, for $|j_r - j_r'| < \log_2(m) + 1$ second compression ideas can never be applied. On coarse levels the number of intersecting quarklets is of order $2^{|j-j'|}$, on higher levels the number of quarklets intersecting the singular support of $\psi_{\lambda'}$ is of order 1. Accordingly splitting the sum leads to

$$
N \lesssim \sum_{\substack{\boldsymbol{p}' \in \mathbb{N}_0^d \\ \prod_{i=1}^d 1 + |p_i - p_i'| \le 2^{\frac{J_1}{a}}}} \sum_{l=\beta_2}^{\beta_1 - 1} \sum_{\substack{\boldsymbol{j}' \in \mathbb{N}_0^d \\ |j-j'|=l}} 1
$$

$$
+ \sum_{\substack{\boldsymbol{p}' \in \mathbb{N}_0^d \\ \prod_{i=1}^d 1 + |p_i - p_i'| \le 2^{\frac{J_1}{a}}}} \sum_{l=0}^{d(\log_2(m)+1)} \sum_{\substack{\boldsymbol{j}' \in \mathbb{N}_0^d \\ |j-j'|=l}} 2^{|j-j'|}.
$$

For the first sum we use $|\{\boldsymbol{j} \in \mathbb{N}_0^d : |\boldsymbol{j}| = l\}| = \binom{l+d-1}{l} \lesssim (1+l)^{d-1}$ to estimate

$$
\sum_{\substack{\boldsymbol{p}' \in \mathbb{N}_0^d \\ \prod_{i=1}^d 1 + |p_i - p_i'| \le 2^{\frac{J_1}{a}}}} \sum_{l=\beta_2}^{\beta_1 - 1} \sum_{\substack{\boldsymbol{j}' \in \mathbb{N}_0^d \\ |j-j'|=l}} 1 \lesssim \sum_{\substack{\boldsymbol{p}' \in \mathbb{N}_0^d \\ \prod_{i=1}^d 1 + |p_i - p_i'| \le 2^{\frac{J_1}{a}}}} \sum_{l=\beta_2}^{\beta_1 - 1} (1+l)^{d-1}
$$

$$
\lesssim \sum_{\substack{\boldsymbol{p}' \in \mathbb{N}_0^d \\ \prod_{i=1}^d 1 + |p_i - p_i'| \le 2^{\frac{J_1}{a}}}} (\beta_1 - \beta_2)^{d-1}.
$$

We further calculate

$$
\sum_{\substack{\boldsymbol{p}'' \in \mathbb{N}^d \\ \prod_{i=1}^d p_i'' \le 2^{\frac{J_1}{a}}}} 1 = \sum_{\substack{\boldsymbol{p}'' \in \mathbb{N}^{d-1} \\ \prod_{i=1}^{d-1} p_i'' \le 2^{\frac{J_1}{a}}}} \sum_{p_d''=1}^{2^{\frac{J_1}{a}}\left(\prod_{i=1}^{d-1} p_i''\right)^{-1}} 1 = \sum_{\substack{\boldsymbol{p}'' \in \mathbb{N}^{d-1} \\ \prod_{i=1}^{d-1} p_i'' \le 2^{\frac{J_1}{a}}}} 2^{J_1/a} \left(\prod_{i=1}^{d-1} p_i''\right)^{-1}.
$$

Similar estimates as in (6.3.15) lead to

$$
\sum_{\substack{\boldsymbol{p}'' \in \mathbb{N}^d \\ \prod_{i=1}^d p_i'' \le 2^{\frac{J_1}{a}}}} 1 \lesssim 2^{J_1/a}(1 + \frac{J_1}{a})^{d-1}.
$$

Combining the previous estimates yields

$$
\sum_{\substack{\boldsymbol{p}' \in \mathbb{N}_0^d \\ \prod_{i=1}^d 1 + |p_i - p_i'| \le 2^{\frac{J_1}{a}}}} \sum_{l=\beta_2}^{\beta_1 - 1} \sum_{\substack{\boldsymbol{j}' \in \mathbb{N}_0^d \\ |j-j'|=l}} 1 \lesssim 2^{J_1/a}(1 + \frac{J_1}{a})^{d-1}(\frac{J_2}{b})^{d-1}.
$$

Similar computations for second sum give an upper bound of the form $2^{J_1/a}(1+\frac{J_1}{a})^{d-1}$, where the inner sums are bounded by a constant. In total, considering also the classical compression case, the matrix A_J has

$$\left(J_1^{2d-2}2^{\frac{J_1}{a}} + J_2^{d-1}2^{\frac{J_2}{b}}\right)\begin{cases} J_2, & a = b, \\ 1, & \text{otherwise.} \end{cases}$$

entries per row and column. Now we turn to (6.3.28). The asymptotic error for $a\log_2(\prod_{i=1}^d 1 + |p_i - p_i'|) + b|\boldsymbol{j} - \boldsymbol{j}'| > J_1$ is given by (6.3.11), hence we just have to consider the case $J_1 \geq a\log_2(\prod_{i=1}^d 1 + |p_i - p_i'|) + b|\boldsymbol{j} - \boldsymbol{j}'| > J_2$, where singular supports intersect in $0, \ldots, d-1$ directions. Again we use the Schur lemma, i.e., we consider

$$\alpha_\lambda := w_\lambda^{-1} \sum_{\lambda' \in \boldsymbol{\nabla}_\sigma} |(\mathbf{A})_{\lambda,\lambda'} - (\mathbf{A}_J)_{\lambda,\lambda'}| w_{\lambda'}.$$

Separating the level and polynomial degree in (6.3.26) and with $\beta_n := \lceil b^{-1}(J_i - a\log_2(\prod_{i=1}^d 1 + |p_i - p_i'|))\rceil, n = 1, 2$ we obtain

$$\alpha_\lambda \lesssim \sum_{\substack{p' \in \mathbb{N}_0^d \\ \prod_{i=1}^d 1+|p_i-p_i'|\leq 2^{J_1/a}}} \sum_{l=\beta_2}^{\beta_1} \sum_{\substack{j':|j-j'|=l \\ \exists r:|j_r-j_r'|>\log(m)+1}} \sum_{\substack{k' \in \boldsymbol{\nabla}_{j',\sigma} \\ \exists r:\lambda_r' \in \bigcup_{j_r',i} A_{j_r',\lambda_r,i}}} 2^{|j|/2}|a_{\lambda,\lambda'}|2^{-|j'|/2},$$

where we chose weights of the form $w_\lambda = 2^{-|j|/2}$. With (6.3.8) we get

$$\alpha_\lambda \lesssim \sum_{\substack{p' \in \mathbb{N}_0^d \\ \prod_{i=1}^d 1+|p_i-p_i'|\leq 2^{J_1/a}}} \left(\prod_{r=1}^d \left(1 + |p_r - p_r'|\right)\right)^{m-1-\delta_2/2} \sum_{l=\beta_2}^{\beta_1} 2^{-l(m\ 2)}$$

$$\lesssim \sum_{\substack{p' \in \mathbb{N}_0^d \\ \prod_{i=1}^d 1+|p_i-p_i'|\leq 2^{J_1/a}}} \left(\prod_{r=1}^d \left(1 + |p_r - p_r'|\right)\right)^{m-1-\delta_2/2} 2^{-\beta_2(m-2)}$$

$$\lesssim 2^{-\frac{J_2}{b}(m-2)} \sum_{\substack{p' \in \mathbb{N}_0^d \\ \prod_{i=1}^d 1+|p_i-p_i'|\leq 2^{J_1/a}}} \left(\prod_{r=1}^d \left(1 + |p_r - p_r'|\right)\right)^{m-1-\delta_2/2+(m-2)\frac{a}{b}}.$$

Similar estimates as in (6.3.15) imply

$$\alpha_\lambda \lesssim 2^{-(m-2)\frac{J_2}{b}}\begin{cases} 2^{(1+\tau+(m-2)\frac{a}{b})J_1/a}(1+\frac{J_1}{a})^{d-1}, & \tau + (m-2)\frac{a}{b} > -1, \\ (1+\frac{J_1}{a})^d, & \tau + (m-2)\frac{a}{b} = -1, \\ 1, & \tau + (m-2)\frac{a}{b} < -1, \end{cases}$$

$$\lesssim \left(2^{-(m-2)\frac{J_2}{b}} + \left(1 + \frac{J_1}{a}\right)^{d-1}2^{(1+\tau)\frac{J_1}{a}}\right)$$

$$\cdot \begin{cases} (1+\frac{J_1}{a}), & \tau + (m-2)\frac{a}{b} = -1, \\ 1, & \text{otherwise.} \end{cases} \qquad \qquad \square$$

Table 6.1: Theoretical comparison of compression techniques.

	Classical compression Theorem 6.15	Second compression Theorem 6.17	Second compression Theorem 6.18
δ_2	$6 + 2\frac{a}{b}$	$10 + 2\frac{a}{b}$	$6 + \frac{a}{b}$
s^*	$\min\{a,b\}/b$	$\min\{a,b\}/b$	$\min\{a,b\}/2b$
$\#$	$2^{J/a} + 2^{J/b}$	$2^{J_1/a} + 2^{J_1/b}$	$2^{J_1/a} + 2^{J_2/b}$

To sum up, we have seen that in the multivariate case additional difficulties come into play, namely the tensor product structure and hence the estimation of inner products of quarklets instead of their derivatives. This circumstances led to less flexibility in the choice of the parameter J_2, in the case of Theorem 6.17 we can only choose $J_2 = \frac{1}{2}J_1$. With $\delta_2 > 2m + 4 + 2\frac{a}{b}$ the rate of compressibility from Theorem 6.15 can be preserved. In the case of Theorem 6.18, i.e., strong second compression, we always lose compressibility, e.g., the value of s^* is halved for $m = 3$, see Table 6.1. On the other hand, the resulting stiffness matrices are much more quasi-sparse. So, strong second compression strategies have to be considered rather as complete different strategies than as an amendment of classical compression strategies.

Chapter 7

Adaptive Quarklet Approximation

In this chapter we discuss approximation properties of quarklet frames for certain univariate singularity functions. In general, the adaptive schemes for the treatment of operator equations can not deliver a better approximation to the exact solution as a direct approach with a given function. On the other hand, for wavelet schemes it is well-known that they realise the asymptotic rate of the best N-term approximation, therefore we study direct approximations. Considering the problem (1.2.4) on smooth domains, the Sobolev regularity of the solution and hence the convergence of uniform schemes is only limited by the regularity of the right-hand-side. The situation on nonsmooth domains, in particular polygonal domains with reentrant corners, is different. On such domains, the Sobolev regularity of the solution is limited due to the interior angle, while the Besov regularity is arbitrarily high and adaptive schemes perform well. Characteristic singularities in higher dimensions are modelled by the univariate function

$$u_\alpha : [0, 1] \to \mathbb{R},$$
$$x \mapsto x^\alpha, \tag{7.1}$$

with $\alpha > \frac{1}{2}$, cf. [2–5].We show that the function u_α can be approximated in L_2 and in H^1, respectively, with exponential order by the elements of our quarklet frame. We proceed in the following way. First of all, we choose a highly nonuniform partition of the interval $[0, 1]$ and approximate u_α by means of a Hermite spline with respect to the partition. Then we show that this spline can be written as a linear combination of quarks on different refinement levels. Finally we rewrite the spline in terms of quarklets and count the necessary degrees of freedom. To this end we employ the results from Section 4.2. It turns out that the result stated in Theorem 7.6 can further be improved. We adapt the ideas from [63] to obtain an approximation with improved error decay, see Theorem 7.9. In the multivariate setting we used a tensor product ansatz to construct quarklet frames. With this at hand, the findings from Sections 7.1 and 7.2 can easily be transferred to the case of edge singularities in higher dimensions in Section 7.3. Let us mention that parts of this chapter have been published in [31].

7.1 Approximation in L_2

7.1.1 Construction of the Spline

We study approximations to (7.1) in terms of quarks and quarklets. First we are going to construct a piecewise polynomial approximation by Hermite interpolation. This generalises the spline from [2]. Thus, for $i = 1, \ldots, J$ we define

- a finite geometric sequence of points: $x_0 := 0$, $x_i := 2^{i-J}$;

- intervals $I_i := [x_{i-1}, x_i]$;

- a local maximal refinement level $j_i := J - i + 1 + \lceil \log_2(m) \rceil$;

- a local maximal polynomial degree $p_i := i + m - 3$.

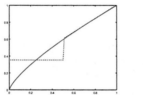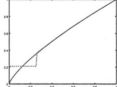

Figure 7.1: Blue: x^α, Red: Variable knot/degree spline for $m = 1$, $J \in \{2, 3\}$.

Theorem 7.1. *Let g be the piecewise polynomial on $[0, 1]$ which on each subinterval I_i is defined as the Hermite interpolant g_i with respect to*

$$\underbrace{x_{i-1}, \ldots, x_{i-1}}_{m-1\ times}, \underbrace{y_i, \ldots, y_i}_{i-1\ times}, \underbrace{x_i, \ldots, x_i}_{m-1\ times}, \tag{7.1.1}$$

where $y_i := \frac{1}{2}(x_i + x_{i-1})$. Let

$$E_i := \|u_\alpha - g_i\|_{L_2(I_i)}^2 \tag{7.1.2}$$

denote the squared L_2 approximation error. Then, it holds that

$$E_i \lesssim 2^{-2i} 2^{(i-J)(2\alpha+1)}, \quad i = 2, \ldots, J, \tag{7.1.3}$$
$$E_1 \lesssim 2^{(1-J)(2\alpha+1)}. \tag{7.1.4}$$

Furthermore the squared global L_2 error is bounded by

$$\sum_{i=1}^{J} E_i \lesssim 2^{-2J}. \tag{7.1.5}$$

Proof. Let $i \geq 2$. We construct g_i as the Hermite interpolant with respect to the $i + 2m - 3$ knots from (7.1.1). For $x \in I_I$ the pointwise error can be estimated by

$$|u_\alpha(x) - g_i(x)| \leq \frac{|u_\alpha^{(i+2m-3)}(\xi)|}{(i+2m-3)!} |x - x_{i-1}|^{m-1} |x - y_i|^{i-1} |x - x_i|^{m-1}$$

$$\leq \frac{|u_\alpha^{(i+2m-3)}(\xi)|}{(i+2m-3)!} |I_i|^{i+2m-3} 2^{-i+1}$$

$$= \frac{|u_\alpha^{(i+2m-3)}(\xi)|}{(i+2m-3)!} 2^{(i-J-1)(i+2m-3)} 2^{-i+1}.$$

Now we estimate the derivative for $\xi \in I_i$. With the absolute convergence of the binomial series it can be bounded by

$$\frac{|u_\alpha^{(i+2m-3)}(\xi)|}{(i+2m-3)!} = \frac{\alpha |\alpha - 1| \cdots |\alpha - i - 2m + 4|}{(i+2m-3)!} \xi^{\alpha - i - 2m + 3} \leq 2^\alpha \xi^{\alpha - i - 2m + 3}.$$

Combining these estimates with the monotonicity of u_α leads to

$$|u_\alpha(x) - g_i(x)| \leq 2^\alpha 2^{(i-J-1)(i+2m-3)} 2^{-i+1} \begin{cases} x_{i-1}^{\alpha-i-2m+3}, & \alpha < i + 2m - 3 \\ x_i^{\alpha-i-2m+3}, & \alpha > i + 2m - 3 \end{cases}$$

$$\leq 2^\alpha \begin{cases} 2^{(i-J-1)\alpha} 2^{-i+1}, & \alpha < i + 2m - 3 \\ 2^{(i-J)\alpha} 2^{-2i-2m+3}, & \alpha > i + 2m - 3 \end{cases}$$

$$\lesssim 2^{(i-J)\alpha} 2^{-i},$$

with a constant depending on m and α. For $E_i, i = 2, \ldots, J$ we conclude

$$E_i = \int_{I_i} |u_\alpha(x) - g_i(x)|^2 dx \lesssim 2^{i-J-1} 2^{2(i-J)\alpha} 2^{-2i} \lesssim 2^{(i-J)(2\alpha+1)} 2^{-2i}.$$

It remains to treat the case $i = 1$. We directly compute

$$E_1 \leq \|u_\alpha\|_{L_2(I_1)}^2 = \int_0^{2^{1-J}} |x^\alpha|^2 dx = \left[\frac{1}{2\alpha+1} x^{2\alpha+1} \right]_0^{2^{1-J}} \leq 2^{(2\alpha+1)(1-J)}.$$

Finally we consider the global error:

$$\sum_{i=2}^J E_i \lesssim \sum_{i=2}^J 2^{-2i} 2^{(2\alpha+1)(i-J)} = 2^{-(2\alpha+1)J} \sum_{i=2}^J \left(2^{-2} 2^{2\alpha+1} \right)^i$$

$$= 2^{-(2\alpha+1)J} \left(2^{-2} 2^{2\alpha+1} \right)^2 \sum_{i=0}^{J-2} \left(2^{-2} 2^{2\alpha+1} \right)^i$$

$$\lesssim 2^{-(2\alpha+1)J} \left(2^{-2} 2^{2\alpha+1} \right)^2 (2^{-2} 2^{2\alpha+1})^{J-1}$$

$$\lesssim 2^{-2J}.$$

With the asymptotic behaviour of E_1 the claim follows. $\qquad\square$

7.1.2 Quarkonial Decomposition

The next step is to show that the approximation $\sum_{i=1}^{J} g_i \chi_{I_i}$ can be expanded in terms of quarklet frame elements. Firstly, we consider a decomposition into fine quarks which are not elements of the frame. Secondly, we use the reconstruction properties derived in Section 4.2 to get a decomposition into frame elements.

Proposition 7.2. *The functions $\varphi_0(\cdot - k), \ldots, \varphi_p(\cdot - k)$, $|k| < m$, span a spline space that contains the polynomial space $\Pi_{p+m-1}(-\lfloor \frac{m}{2} \rfloor, \lceil \frac{m}{2} \rceil)$.*

Proof. Let $q \in \mathbb{N}_0$. If $q \le m - 1$, x^q has a representation in terms of B-splines, i.e., for $x \in (-\lfloor \frac{m}{2} \rfloor, \lceil \frac{m}{2} \rceil)$ it holds

$$x^q = \sum_{-m < k < m} c_{q,k} \varphi_0(x - k).$$

Otherwise we split $q = p + m - 1$ such that

$$
\begin{aligned}
x^q = x^p x^{m-1} &= x^p \sum_{-m < k < m} c_{m-1,k} \varphi_0(x - k) \\
&= \sum_{-m < k < m} \sum_{l=0}^{p} \binom{p}{l} k^{p-l} (x-k)^l c_{m-1,k} \varphi_0(x - k) \\
&= \sum_{-m < k < m} \sum_{l=0}^{p} \binom{p}{l} \lceil m/2 \rceil^l k^{p-l} c_{m-1,k} \varphi_l(x - k).
\end{aligned}
\tag{7.1.6}
$$

$\square$

By a change of variables, we can expand any polynomial on the intervals I_i in terms of quark generators:

Proposition 7.3. *The quarks $\varphi_{l,j_i,k}, l = 0, \ldots, p_i$, with translation parameters $2^{\lceil \log_2(m) \rceil} - \lceil m/2 \rceil < k < 2^{\lceil \log_2(m) \rceil + 1} + \lfloor m/2 \rfloor$ span a spline space including the polynomial space $\Pi_{i+2m-4}(I_i)$.*

Proof. We have $\varphi_{p,j_i,k} = 2^{j_i/2} \varphi_p(2^{j_i} \cdot - k)$ and hence

$$\operatorname{supp} \varphi_{p,j_i,k} = [2^{-j_i}(k - \lfloor m/2 \rfloor), 2^{-j_i}(k + \lceil m/2 \rceil)]. \tag{7.1.7}$$

Comparing the interval bounds with $I_i = [2^{i-J-1}, 2^{i-J}]$, we conclude that only the supports of those quarks intersect with I_i which satisfy

$$2^{\lceil \log_2(m) \rceil} - \lceil m/2 \rceil < k < 2^{\lceil \log_2(m) \rceil + 1} + \lfloor m/2 \rfloor. \tag{7.1.8}$$

For quarks lying completely inside I_i the condition on the translation parameter reads as follows:

$$2^{\lceil \log_2(m) \rceil} + \lfloor m/2 \rfloor \le k \le 2^{\lceil \log_2(m) \rceil + 1} - \lceil m/2 \rceil. \tag{7.1.9}$$

Hence, the B-splines fulfilling (7.1.8) generate all polynomials with degree $m - 1$ on I_i. The remainder of the proof is analogous to the previous one. $\square$

So far, we have shown that each g_i on I_i can be decomposed in terms of quarks on level j_i. But, for $m \geq 2$ the supports of certain quarks intersect with the intervals I_{i-1}, I_{i+1}. As a consequence we have to thin out the amount of quarks in the neighbourhood of the knots x_i to construct a globally $m-1$-times differentiable approximation to u_α. We do this in the following way. We allow the supports of the quarks $\varphi_{p,j_i,k}$ to intersect with the interval I_{i+1}, but not with I_{i-1}. At the left boundary of I_i, for each p we insert $m-1$ quarks on level $j_i - 1$. By proceeding this way and using the refinability of the B-splines the gap in the spline space of degree $m-1$ is filled. In the following we use the abbreviations

$$K_0 := 2^{\lceil \log_2(m) \rceil}, \quad K_1 := 2^{\lceil \log_2(m) \rceil + 1}. \tag{7.1.10}$$

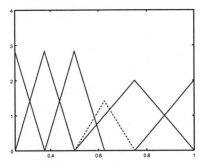

Figure 7.2: Closing the 'gap' in the spline space at the right-hand-side of $x_i = \frac{1}{2}$ with an additional fine hat.

Proposition 7.4. *Let $2 \leq i < J$, $j_i = J - i + 1 + \lceil \log_2(m) \rceil$ and $p_i = i + m - 3$. Then, the coarse quarks $\varphi_{p,j_i,k}, p = 0, \ldots, p_i$, $K_0 + \lfloor m/2 \rfloor \leq k < K_1 + \lfloor m/2 \rfloor$ and the fine quarks $\varphi_{p,j_i+1,k}, p = 0, \ldots, p_i - 1, K_1 - \lceil m/2 \rceil < k \leq K_1 + \lfloor m/2 \rfloor + m - 2$ span a spline space including the polynomial space $\Pi_{i+2m-5}(I_i)$.*

Proof. As seen in the proof of Proposition 7.2, it suffices to show that each polynomial of degree $m-1$ can be decomposed into a sum of B-splines on fine and coarse scales. Let $P \in \Pi_{m-1}(I_i)$. Of course, P has a decomposition in terms of coarse B-splines

$$P(x) = \sum_{K_0 - \lceil m/2 \rceil < k < K_1 + \lfloor m/2 \rfloor} c_k \varphi_{0,j_i,k}(x)$$

$$= \sum_{k=K_0 - \lceil m/2 \rceil + 1}^{K_0 + \lfloor m/2 \rfloor - 1} c_k \varphi_{0,j_i,k}(x) + \sum_{k=K_0 + \lfloor m/2 \rfloor}^{K_1 + \lfloor m/2 \rfloor - 1} c_k \varphi_{0,j_i,k}(x),$$

where the latter sum consists only of B-splines not intersecting with I_{i-1}. We have a look at the first sum. Inserting the refinement equation for B-splines (4.1.5) yields

$$\sum_{k=K_0 - \lceil m/2 \rceil + 1}^{K_0 + \lfloor m/2 \rfloor - 1} c_k \varphi_0(2^{j_i} x - k) = \sum_{k=K_0 - \lceil m/2 \rceil + 1}^{K_0 + \lfloor m/2 \rfloor - 1} c_k \left(\sum_{l=-\lfloor m/2 \rfloor}^{\lceil m/2 \rceil} a_l \varphi_0(2 \cdot 2^{j_i} x - 2k - l) \right).$$

131

With an index shift we obtain

$$\sum_{k=K_0-\lceil m/2\rceil+1}^{K_0+\lfloor m/2\rfloor-1} c_k\varphi_0(2^{j_i}x-k) = \sum_{\bar{l}=K_1-m-\lceil m/2\rceil+2}^{K_1+m+\lfloor m/2\rfloor-2} \left(\sum_{2k+l=\bar{l}} c_k a_l\right)\varphi_0(2^{j_i+1}x-\bar{l}).$$

With (7.1.9) we can omit the fine B-splines lying completely inside I_{i-1} and get

$$\sum_{k=K_0-\lceil m/2\rceil+1}^{K_0+\lfloor m/2\rfloor-1} c_k\varphi_0(2^{j_i}x-k) = \sum_{\bar{l}=K_1-\lceil m/2\rceil+1}^{K_1+m+\lfloor m/2\rfloor-2} \left(\sum_{2k+l=\bar{l}} c_k a_l\right)\varphi_0(2^{j_i+1}x-\bar{l}).$$

Hence we have for $x \in I_i$:

$$P(x) = \sum_{\bar{l}=K_1-\lceil m/2\rceil+1}^{K_1+\lfloor m/2\rfloor+m-2} \left(\sum_{2k+l=\bar{l}} c_k a_l\right)\varphi_{0,j_i+1,\bar{l}}(x)$$

$$+ \sum_{k=K_0+\lfloor m/2\rfloor}^{K_1+\lfloor m/2\rfloor-1} c_k\varphi_{0,j_i,k}(x).$$

By polynomial enrichment as in (7.1.6), the claim follows. $\square$

Theorem 7.5. *Let the spline constructed in Theorem 7.1 be given by $g = \sum_{i=1}^{J} g_i \chi_{I_i}$ and let j_i be defined as at the beginning of this section. We define the collection of quark indices*

$$\Lambda_{j_i} := \{(p,j_i,k) : p \leq p_i, K_0 + \lfloor m/2\rfloor \leq k \leq K_1 + \lfloor m/2\rfloor + m - 2\},$$
$$\Lambda_{j_1} := \{(p,j_1,k) : p \leq p_1, \lfloor m/2\rfloor \leq k \leq K_1 + \lfloor m/2\rfloor + m - 2\},$$
$$\Lambda_{j_J} := \{(p,j_J,k) : p \leq p_J, K_0 + \lfloor m/2\rfloor \leq k < K_1 + \lfloor m/2\rfloor\},$$
$$\Lambda := \bigcup_{i=1}^{J} \Lambda_{j_i}. \tag{7.1.11}$$

Then, there exist $c_\lambda \in \mathbb{R}$, such that

$$g(x) = \sum_{\lambda\in\Lambda} c_\lambda \varphi_\lambda(x), \quad x \in I. \tag{7.1.12}$$

Proof. Let $2 \leq i \leq J$. We proceed in the following way: Suppose, that g_{i-1} is given on I_{i-1} as a polynomial of degree $i + 2m - 5$. Then, we decompose g_i into

$$g_i(x) = P_i(x) + Q_i(x),$$

where $P_i \in \Pi_{i+2m-5}$, $Q_i \in \Pi_{i+2m-4}$ and Q_i has a root with multiplicity $m - 1$ in x_{i-1}. P_i is constructed as the extension of $g_{i-1}|_{I_i}$ to a polynomial on I_i. With (7.1.8) and (7.1.9) we can rewrite Λ as $\Lambda = \cup\Lambda_{I_i}$,

$$\Lambda_{I_i} := \{(p,j_{i-1},k) : p \leq p_{i-1}, K_1 - \lceil m/2\rceil < k < K_1 + \lfloor m/2\rfloor\}$$
$$\cup\{(p,j_{i-1},k) : p \leq p_{i-1}, K_1 + \lfloor m/2\rfloor \leq k \leq K_1 + \lfloor m/2\rfloor + m - 2\}$$
$$\cup\{(p,j_i,k) : p \leq p_i, K_0 + \lfloor m/2\rfloor \leq k < K_1 + \lfloor m/2\rfloor\}.$$

Since g_{i-1} is a polynomial on I_{i-1}, with Proposition 7.3 we have for $x \in I_{i-1}$:

$$g_{i-1}(x) = \sum_{\lambda \in \Lambda_{I_{i-1}}} c_\lambda \varphi_\lambda(x).$$

Considering g_{i-1} on I_i, we can omit the nonoverlapping quarks and get for $x \in I_i$

$$g_{i-1}|_{I_i}(x) = \sum_{k=K_1-\lceil m/2 \rceil+1}^{K_1+\lfloor m/2 \rfloor-1} \sum_{p=0}^{p_i-1} c_{k,p} \varphi_{p,j_i+1,k}(x).$$

With Proposition 7.4, $g_{i-1}|_{I_i}$ can be extended to a polynomial $P_i \in \Pi_{i+2m-5}(I_i)$:

$$P_i(x) = \sum_{k=K_1-\lceil m/2 \rceil+1}^{K_1+\lfloor m/2 \rfloor+m-2} \sum_{p=0}^{p_i-1} c_{k,p} \varphi_{p,j_i+1,k}(x) + \sum_{k=K_0+\lfloor m/2 \rfloor}^{K_1+\lfloor m/2 \rfloor-1} \sum_{p=0}^{p_i-1} c_{k,p} \varphi_{p,j_i,k}(x).$$

By construction, the polynomial Q_i interpolates $u_\alpha - P_i$ in x_i and y_i and has a decomposition

$$Q_i(x) = (x - x_{i-1})^{m-1} \sum_{p=0}^{p_i} a_p x^p.$$

Obviously, the first part consists only of B-splines lying completely inside I_i. Hence, Q_i has a decomposition in quarks on a coarse level:

$$Q_i(x) = \sum_{k=K_0+\lfloor m/2 \rfloor}^{K_1+\lfloor m/2 \rfloor-1} \sum_{p=0}^{p_i} b_{k,p} \varphi_{p,j_i,k}(x).$$

The case $i = 1$ is already covered by Proposition 7.3, since no overlapping quarks have to be considered. $\qquad \square$

After all these preparations, we are now able to state and to prove the main result of this section. By expanding the Hermite interpolation spline with respect to the elements of the quarklet frame, we show that the model function u_α can indeed be approximated with exponential order.

Theorem 7.6. *Let* $\nabla := \{(p,j,k) : p \in \mathbb{N}_0, j \in \mathbb{N}_0 \cup \{-1\}, k \in \mathbb{Z}\}$ *be the index set of the full quarklet system and let* g *be the spline constructed in Theorem 7.1. For* $N \sim J^5$, $J \in \mathbb{N}$ *there exist* $c_\lambda \in \mathbb{R}$ *such that*

$$g(x) = \sum_{\lambda \in \nabla' \subset \nabla : |\nabla'| \leq N} c_\lambda \psi_\lambda(x), \quad x \in I, \tag{7.1.13}$$

$$\|u_\alpha - g\|_{L_2(I)}^2 \lesssim \left(2^2\right)^{-N^{1/5}} = e^{-2\ln(2)N^{1/5}}. \tag{7.1.14}$$

Proof. First we have a look at (7.1.13). Since we have a decomposition of g in terms of quarks, see (7.1.12), and finite reconstruction sequences derived in Section 4.2, (7.1.13) follows. It remains to estimate the asymptotic number of frame elements. With Proposition 7.4, each polynomial p_i on I_i has a decomposition in terms of $C_0(m)p_i$ fine quarks on level j_i. With Theorem 4.18, each fine multiquark consists of $j_i^2 p_i$ frame element vectors. Hence, summation over i gives a total number of J^5 frame elements. Inserting this into the estimate (7.1.5) gives (7.1.14). $\qquad \square$

7.1.3 Quarkonial Decomposition: An Improved Approach

The previously discussed approach has certain advantages. By employing the general machinery from Chapter 3, it is ready for generalisation to other multiwavelet systems. Furthermore it is constructive in the sense that the auxiliary fine quarks can easily be developed in terms of frame elements. However, it is possible to improve the approximation rate to $e^{-\sqrt[3]{N}}$ by using a different approach. In this section we pursue this approach which was originally developed in [63]. We do not rely on the reconstruction properties of the quarklets derived in Section 4.2, but make use of the original reconstruction properties of the wavelets.

We begin with recalling some notation. Let $\varphi = N_m(\cdot + \lfloor \frac{m}{2} \rfloor)$ denote the symmetrised cardinal B-spline of order $m \in \mathbb{N}$, see (2.1.34). Let the cardinal B-spline wavelet ψ by given by its two scale relation (2.1.25) with mask $b = \{b_l\}_{l\in\mathbb{Z}}$. As usual we consider translated and scaled versions for $j \in \mathbb{Z}, j \geq 0, k \in \mathbb{Z}$:

$$\varphi_{j,k} := 2^{j/2}\varphi(2^j \cdot -k), \tag{7.1.15}$$

$$\psi_{j,k} := 2^{j/2}\psi(2^j \cdot -k). \tag{7.1.16}$$

We omit the scaling factor from Section 4.1 and define quarks and quarklets by

$$\varphi_p := (\cdot)^p\varphi, \tag{7.1.17}$$

$$\varphi_{p,j,k} := 2^{j/2}\varphi_p(2^j \cdot -k). \tag{7.1.18}$$

Additionally we define auxiliary quarks and quarklets by

$$\widetilde{\varphi}_p := (\cdot)^p\varphi, \tag{7.1.19}$$

$$\widetilde{\varphi}_{p,j,k} := (\cdot)^p\varphi_{j,k}, \tag{7.1.20}$$

$$\widetilde{\psi}_p := (\cdot)^p\psi, \tag{7.1.21}$$

$$\widetilde{\psi}_{p,j,k} := (\cdot)^p\psi_{j,k}. \tag{7.1.22}$$

The technique developed in [63] heavily uses a beneficial alternative to the reconstruction formula (4.2.13). Instead of using reconstruction sequences for multiquarks, we rely on the reconstruction sequences associated with the underlying wavelet basis.

Proposition 7.7. *Let $p \in \mathbb{N}_0$, $j \geq j_0, k \in \mathbb{Z}$. For each $\widetilde{\varphi}_{p,j+1,k}$ there exist sequences $c_{j,k} = \{c_{j,k,l}\}_{l\in\mathbb{Z}}, d_{j,k} = \{d_{j,k,n}\}_{n\in\mathbb{Z}}$ such that $\widetilde{\varphi}_{p,j+1,k}$ has a decomposition into functions $\widetilde{\varphi}_{p,j,l}, \widetilde{\psi}_{p,j,n}$ of the form*

$$\widetilde{\varphi}_{p,j+1,k} = \sum_{l\in\mathbb{Z}} c_{j,k,l}\widetilde{\varphi}_{p,j,l} + \sum_{n\in\mathbb{Z}} d_{j,k,n}\widetilde{\psi}_{p,j,n}. \tag{7.1.23}$$

Proof. Rewriting (2.1.31), the underlying B-spline wavelet basis fulfils the following reconstruction property. For each $\varphi_{j,k}$ there exist sequences $c_{j,k} = \{c_{j,k,l}\}_{l\in\mathbb{Z}}$, $d_{j,k} = \{d_{j,k,n}\}_{n\in\mathbb{Z}}$ such that

$$\varphi_{j+1,k} = \sum_{l\in\mathbb{Z}} c_{j,k,l}\varphi_{j,l} + \sum_{n\in\mathbb{Z}} d_{j,k,n}\psi_{j,n}. \tag{7.1.24}$$

Note that the reconstruction sequences are finitely supported. Applying the polynomial enrichment from (7.1.20), (7.1.22) immediately gives the claim. □

Next we study how the quarks $\varphi_{p,j,k}$ can be transformed into functions $\widetilde{\varphi}_{p,j,k}$.

Proposition 7.8. *Let $\{\lambda_{p,j,k}\}_{k\in\mathbb{Z}} \in \ell_1(\mathbb{Z})$. For linear combination of quarks it holds that*

$$\sum_{k\in\mathbb{Z}} \lambda_{p,j,k}\varphi_{p,j,k} = \sum_{k\in\mathbb{Z}} 2^{jp}\lambda_{p,j,k}\widetilde{\varphi}_{p,j,k} + \sum_{q=0}^{p-1}\sum_{k\in\mathbb{Z}} \mu_{q,j,k}\widetilde{\varphi}_{q,j,k}, \qquad (7.1.25)$$

where $\operatorname{supp}\mu_{q,j} = \operatorname{supp}\lambda_{p,j}$ *for all* $0 \le q \le p$.

Proof. With the above definitions we have

$$\begin{aligned}
\varphi_{p,j,k} &:= 2^{j/2}\varphi_p(2^j \cdot -k) \\
&= 2^{j/2}(2^j \cdot -k)^p\varphi(2^j \cdot -k) \\
&= (2^j \cdot -k)^p\varphi_{j,k}.
\end{aligned}$$

By the binomial theorem we calculate

$$\begin{aligned}
\sum_{k\in\mathbb{Z}} \lambda_{p,j,k}\varphi_{p,j,k} &= \sum_{k\in\mathbb{Z}} \lambda_{p,j,k}(2^j \cdot -k)^p\varphi_{j,k} \\
&= \sum_{k\in\mathbb{Z}} \lambda_{p,j,k}\left(\sum_{q=0}^{p}\binom{p}{q}(2^j\cdot)^q(-k)^{p-q}\right)\varphi_{j,k} \\
&= \sum_{k\in\mathbb{Z}} \lambda_{p,j,k}2^{jp}(\cdot)^p\varphi_{j,k} + \sum_{k\in\mathbb{Z}} \lambda_{p,j,k}\left(\sum_{q=0}^{p-1}\binom{p}{q}2^{jq}(-k)^{p-q}\right)(\cdot)^q\varphi_{j,k}.
\end{aligned}$$

With $\mu_{q,j,k} = \lambda_{p,j,k}\binom{p}{q}2^{jq}(-k)^{p-q}$ the claim follows. □

As an immediate consequence of Proposition 7.8 we get the following relation between the quarklets $\psi_{p,j,k}$ and $\widetilde{\psi}_{p,j,k}$:

$$\psi_{p,j,k} = 2^{jp}\widetilde{\psi}_{p,j,k} + \sum_{q=0}^{p-1}\sum_{l\in\mathbb{Z}} \mu_{q,j,l}\widetilde{\varphi}_{q,j+1,l}, \qquad k \in \mathbb{Z}, \qquad (7.1.26)$$

where $\operatorname{supp}\mu_{q,j} = \operatorname{supp}b_k^j$. Combining (7.1.23) and (7.1.26) leads to the relation

$$\widetilde{\varphi}_{p,j+1,k} = \sum_{l\in\mathbb{Z}} c_{p,j,l}\widetilde{\varphi}_{p,j,l} + \sum_{n\in\mathbb{Z}} 2^{jp}d_{p,j,n}\psi_{p,j,n} + \sum_{q=0}^{p-1}\sum_{s\in\mathbb{Z}} \mu_{q,j,s}\widetilde{\varphi}_{q,j+1,s}, \qquad k \in \mathbb{Z}. \quad (7.1.27)$$

Again, we have $\operatorname{supp}\mu_{q,j} = \operatorname{supp}d_{p,j}, 0 \le q < p$. To keep the following calculations as simple as possible, we introduce the following notation. Without loss of generality we assume the translation parameter to begin at $k = 0$. Furthermore, we assume $p_i = i$, $j_i = J - i$ and $j_0 = 0$:

$$\text{span}_{n_0}^{p_1,\dots,p_2,j}\{\widetilde{\varphi}\} := \text{span}\{\widetilde{\varphi}_{q,j,k} : p_1 \leq q \leq p_2, 0 < k < n_0\}. \tag{7.1.28}$$

Now we are able to deduce a recursion for expressing the spline g in Theorem 7.1 in terms of functions $\widetilde{\varphi}_{p,j,k}$ and frame elements $\psi_{p,j,k}$. From (7.1.11) we infer that, on each level and polynomial degree, the number of translations is bounded by

$$n_0 := 2^{\lceil \log_2(m) \rceil} + m - 1. \tag{7.1.29}$$

Since the length of the reconstruction sequences in this section only depends on the underlying wavelet basis, we can fix a number

$$n := \max\{|\operatorname{supp} c_j|, |\operatorname{supp} d_j|\}. \tag{7.1.30}$$

With this at hand and using Proposition 7.8, we can reformulate (7.1.12) as

$$g \in \sum_{i=1}^{J} \text{span}_{n_0}^{0,\dots,i,J-i}\{\widetilde{\varphi}\}. \tag{7.1.31}$$

Separating the generator level leads to

$$\sum_{i=1}^{J} \text{span}_{n_0}^{0,\dots,i,J-i}\{\widetilde{\varphi}\} = \text{span}_{n_0}^{0,\dots,J,0}\{\widetilde{\varphi}\} + \sum_{i=1}^{J-1} \text{span}_{n_0}^{0,\dots,i,J-i}\{\widetilde{\varphi}\}. \tag{7.1.32}$$

Applying (7.1.27) only for the highest polynomial degree i yields

$$\sum_{i=1}^{J} \text{span}_{n_0}^{0,\dots,i,J-i}\{\widetilde{\varphi}\} = \text{span}_{n_0}^{0,\dots,J,0}\{\widetilde{\varphi}\} + \sum_{i=1}^{J-1} \text{span}_{n_0+2n}^{i,\dots,i,J-i-1}\{\widetilde{\varphi}\}$$

$$+ \sum_{i=1}^{J-1} \text{span}_{n_0+2n}^{i,\dots,i,J-i-1}\{\psi\} + \sum_{i=1}^{J-1} \text{span}_{n_0+2n}^{0,\dots,i-1,J-i}\{\widetilde{\varphi}\}.$$

We consider the first sum. Again separating the generator level gives

$$\sum_{i=1}^{J} \text{span}_{n_0}^{0,\dots,i,J-i}\{\widetilde{\varphi}\} = \text{span}_{n_0}^{0,\dots,J,0}\{\widetilde{\varphi}\} + \text{span}_{n_0+2n}^{J-1,\dots,J-1,0}\{\widetilde{\varphi}\}$$

$$+ \sum_{i=1}^{J-2} \text{span}_{n_0+2n}^{i,\dots,i,J-i-1}\{\widetilde{\varphi}\} + \sum_{i=1}^{J-1} \text{span}_{n_0+2n}^{i,\dots,i,J-i-1}\{\psi\}$$

$$+ \sum_{i=1}^{J-1} \text{span}_{n_0+2n}^{0,\dots,i-1,J-i}\{\widetilde{\varphi}\}.$$

Now we turn to the last sum. Separating the case $i = 1$ leads to

$$\sum_{i=1}^{J} \text{span}_{n_0}^{0,\dots,i,J-i}\{\tilde\varphi\} = \text{span}_{n_0}^{0,\dots,J,0}\{\tilde\varphi\} + \text{span}_{n_0+2n}^{J-1,\dots,J-1,0}\{\tilde\varphi\}$$

$$+ \sum_{i=1}^{J-2} \text{span}_{n_0+2n}^{i,\dots,i,J-i-1}\{\tilde\varphi\} + \sum_{i=1}^{J-1} \text{span}_{n_0+2n}^{i,\dots,i,J-i-1}\{\psi\}$$

$$+ \sum_{i=2}^{J-1} \text{span}_{n_0+2n}^{0,\dots,i-1,J-i}\{\tilde\varphi\} + \text{span}_{n_0+2n}^{0,\dots,0,J-1}\{\tilde\varphi\}.$$

Performing an index shift, wee notice that the spans of the fine quarks coincide. Hence we have

$$\sum_{i=1}^{J} \text{span}_{n_0}^{0,\dots,i,J-i}\{\tilde\varphi\} = \text{span}_{n_0}^{0,\dots,J,0}\{\tilde\varphi\} + \text{span}_{n_0+2n}^{J-1,\dots,J-1,0}\{\tilde\varphi\}$$

$$+ \sum_{i=1}^{J-2} \text{span}_{n_0+2n}^{0,\dots,i,J-i-1}\{\tilde\varphi\} + \sum_{i=1}^{J-1} \text{span}_{n_0+2n}^{i,\dots,i,J-i-1}\{\psi\}$$

$$+ \text{span}_{n_0+2n}^{0,\dots,0,J-1}\{\tilde\varphi\}.$$

This procedure can be iterated. Summing up the number of the separated quarklets in each recursion step gives the total amount of used degrees of freedom. In the first step, $(n_0 + 2n)(J - 1) \sim J$ quarklets are separated. In the second step, the recursion (7.1.27) is applied to the $n_0 + 2n$ remaining fine quarks, what leads to a linearly increasing number of quarklets in each step. In the last step, J^2 quarklets are separated. It remains to consider the remaining fine wavelet generators. It is known that the classical reconstruction of a fine generator needs J^2 wavelets. Summing up the quarklets per step leads to a total amount of J^3 frame elements. We get the following improved result.

Theorem 7.9. *Let $\nabla := \{(p, j, k) : p \in \mathbb{N}_0, j \in \mathbb{N}_0 \cup \{-1\}, k \in \mathbb{Z}\}$ be the index set of the full quarklet system and let g be the spline constructed in Theorem 7.1. For $N \sim J^3$, $J \in \mathbb{N}$ there exist $c_\lambda \in \mathbb{R}$ such that*

$$g(x) = \sum_{\lambda \in \nabla' \subset \nabla : |\nabla'| \leq N} c_\lambda \psi_\lambda(x), \quad x \in I, \tag{7.1.33}$$

$$\|u_\alpha - g\|_{L_2(I)}^2 \lesssim \left(2^2\right)^{-N^{1/3}} = e^{-2\ln(2)N^{1/3}}. \tag{7.1.34}$$

7.2 Approximation in H^1

Theorem 7.10. *Let g be the piecewise polynomial on $[0, 1]$ whose derivative on each subinterval I_i is defined as the Hermite interpolant g_i with respect to u'_α and the knots*

$$\underbrace{x_{i-1}, \dots, x_{i-1}}_{m-1-times}, \underbrace{y_i, \dots, y_i}_{i-2-times}, \underbrace{x_i, \dots, x_i}_{m-1-times}, \tag{7.2.1}$$

where $y_i := \frac{1}{2}(x_{i-1} + x_i)$. Let

$$E_i := |u_\alpha - g|^2_{H^1(I_i)} \tag{7.2.2}$$

denote the squared H^1 approximation error. Then, it holds that

$$E_i \lesssim 2^{-2i} 2^{(i-J)(2\alpha-1)}, \quad i = 2, \dots, J, \tag{7.2.3}$$
$$E_1 \lesssim 2^{(1-J)(2\alpha-1)}. \tag{7.2.4}$$

Furthermore the global squared H^1 error is bounded by

$$\sum_{i=1}^{J} E_i \lesssim \min\left(2^2, 2^{2\alpha-1}\right)^{-J}. \tag{7.2.5}$$

Proof. We use

$$|u|_{H^1(I_i)} = \|u'\|_{L_2(I_i)}.$$

Now the remainder of the proof is analogous to the previously treated L_2-case. Let $i \geq 2$ and consider the $i + 2m - 4$ knots in (7.2.1). With $u'_\alpha(x) = \alpha x^{\alpha-1}$ and $g_i \in \Pi_{i+2m-5}$ we conclude for $x \in I_i$

$$
\begin{aligned}
|u'_\alpha(x) - g_i(x)| &\leq \frac{|u'^{(i+2m-4)}_\alpha(\xi)|}{(i+2m-4)!} |x - x_{i-1}|^{m-1} |x - y_i|^{i-2} |x - x_i|^{m-1} \\
&\leq \frac{|u^{(i+2m-3)}_\alpha(\xi)|}{(i+2m-4)!} |I_i|^{i+2m-4} 2^{-i+2} \\
&= \frac{|u^{(i+2m-3)}_\alpha(\xi)|}{(i+2m-4)!} 2^{(i-J-1)(i+2m-4)} 2^{-i+2}.
\end{aligned}
$$

Now we have a look at the derivative for $\xi \in I_i$. By Cauchy's theorem we have

$$\frac{|u'^{(i+2m-4)}_\alpha(\xi)|}{(i+2m-4)!} \leq R^{-i-2m+4} \max_{|\xi-y| \leq R} |u'_\alpha(\xi)|.$$

Choosing $R = 2^{i-1-J}$ und using the monotonicity of u_α leads to

$$
\begin{aligned}
|u'_\alpha(x) - g_i(x)| &\leq \alpha 2^{(i-1-J)(-i-2m+4)} 2^{(i-J-1)(i+2m-4)} 2^{-i+2} \\
&\begin{cases} 2^{(i-2-J)(\alpha-1)}, & \alpha < 1 \\ 5^{\alpha-1} 2^{(i-2-J)(\alpha-1)}, & \alpha > 1 \end{cases} \\
&\lesssim 2^{(i-J)(\alpha-1)} 2^{-i},
\end{aligned}
$$

with a constant depending on m and α. For $E_i, i = 2, \dots, J$ we conclude

$$E_i = \int_{I_i} |u'_\alpha(x) - g_i(x)|^2 dx \lesssim 2^{i-J-1} 2^{2(i-J)(\alpha-1)} 2^{-2i} \lesssim 2^{(i-J)(2\alpha-1)} 2^{-2i}.$$

Now let $i = 1$. We directly compute

$$E_1 \leq \|u'_\alpha\|^2_{L_2(I_1)} = \alpha^2 \int_0^{2^{1-J}} |x^{\alpha-1}|^2 dx = \left[\frac{\alpha^2 x^{2\alpha-1}}{2\alpha-1} \right]_0^{2^{1-J}} \lesssim 2^{(2\alpha-1)(1-J)}.$$

Next we consider the global error:

$$\sum_{i=2}^{J} E_i \lesssim \sum_{i=2}^{J} 2^{-2i} 2^{(2\alpha-1)(i-J)} = 2^{-(2\alpha-1)J} \sum_{i=2}^{J} \left(2^{-2} 2^{2\alpha-1} \right)^i$$

$$= 2^{-(2\alpha-1)J} \left(2^{-2} 2^{2\alpha-1} \right)^2 \sum_{i=0}^{J-2} \left(2^{-2} 2^{2\alpha-1} \right)^i$$

$$\lesssim 2^{-(2\alpha-1)J} \left(2^{-2} 2^{2\alpha-1} \right)^2 (2^{-2} 2^{2\alpha-1})^{J-1}$$

$$\lesssim 2^{-2J}.$$

With the asymptotic behaviour of E_1 the claim follows. $\qquad\square$

Theorem 7.11. *Let $\nabla := \{(p, j, k) : p \in \mathbb{N}_0, j \in \mathbb{N}_0 \cup \{-1\}, k \in \mathbb{Z}\}$ be the index set of the full quarklet system and let g be the spline constructed in Theorem 7.10. For $N \sim J^3$, $J \in \mathbb{N}$ there exist $c_\lambda \in \mathbb{R}$ such that*

$$g(x) = \sum_{\lambda \in \nabla' \subset \nabla : |\nabla'| \leq N} c_\lambda \psi_\lambda(x), \quad x \in I, \tag{7.2.6}$$

$$|u_\alpha - g|^2_{H^1(I)} \lesssim \min \left(2^2, 2^{2\alpha-1} \right)^{-N^{1/3}} = e^{-\min(2, 2\alpha-1)\ln(2)N^{1/3}}. \tag{7.2.7}$$

Proof. To derive a quarkonial decomposition of the polynomial approximation, we differentiate

$$x^{q-1} = \left(\frac{1}{q} x^q \right)' = \sum_\lambda \tilde{c}_\lambda \varphi'_\lambda(x), \quad q = 1, \ldots, i + 2m - 4.$$

That means that every polynomial on I_i can be decomposed with respect to derivatives of quarks. Using the decomposition of quarks on a fine level in terms of frame elements as in Theorem 7.9, we get the asymptotic error decay. $\qquad\square$

7.3 Tensor Product Quarklet Approximation

Let us now consider the case of the unit cube $I^2 = [0, 1]^2$. As a model for edge singularities that might occur in higher dimensions we consider the function

$$\begin{aligned} u_\alpha &: I^2 \to \mathbb{R}, \\ x &\mapsto x_1^\alpha, \end{aligned} \tag{7.3.1}$$

with $\alpha > \frac{1}{2}$. We expect that anisotropic singularities of the form (7.3.1) can be very efficiently approximated by anisotropic tensor product quarklets. This is indeed the case, as we shall see below. For a description of typical isotropic singularities that are located in vertices we refer to [6, 7].

Theorem 7.12. *Let $\tilde{g}$ be the univariate spline constructed in Theorem 7.10. Then we define the function $g(x_1, x_2) := \tilde{g}(x_1)\chi_{[0,1]}(x_2)$. Let*

$$E_i := |u_\alpha - g|^2_{H^1(I_i \times I)}, \tag{7.3.2}$$

denote the squared H^1 approximation error. Then, it holds that

$$E_i \lesssim 2^{-2i} 2^{(i-J)(2\alpha-1)}, \quad i = 2, \ldots, J, \tag{7.3.3}$$
$$E_1 \lesssim 2^{(1-J)(2\alpha-1)}. \tag{7.3.4}$$

Furthermore the global squared H^1 error is bounded by

$$\sum_{i=1}^{J} E_i \lesssim \min\left(2^2, 2^{2\alpha-1}\right)^{-J}. \tag{7.3.5}$$

Proof. The proof is analogous to the univariate case. We use

$$|u|^2_{H^1(I_i \times I)} = \sum_{\beta \in \mathbb{N}^2, |\beta|=1} \|D^\beta u\|_{L_2(I_i \times I)}, \tag{7.3.6}$$

where $D^{(0,1)}u_\alpha = 0$. Hence it suffices to consider derivatives with respect to x_1, i.e., $\frac{\partial}{\partial x_1}u$. We construct the Hermite interpolation polynomial g_i as $g_i(x) := \tilde{g}_i(x_1)$. Following the lines of the proof of Theorem 7.10, we obtain

$$|\frac{\partial}{\partial x_1}u_\alpha(x) - g_i(x)| \lesssim 2^{(i-J)(\alpha-1)} 2^{-i}. \tag{7.3.7}$$

Now we conclude for E_i:

$$E_i = \int_I \int_{I_i} |\frac{\partial}{\partial x_1}u_\alpha(x) - g_i(x)|^2 \mathrm{d}x \lesssim \int_I 2^{i-J-1} 2^{2(i-J)(\alpha-1)} 2^{-2i} \mathrm{d}x_2$$
$$\lesssim 2^{(i-J)(2\alpha-1)} 2^{-2i}.$$

Similar computations for E_1 and summation over i complete the proof. $\square$

Theorem 7.13. *Let $\nabla := \{(p, j, k) : p \in \mathbb{N}_0, j \in \mathbb{N}_0 \cup \{-1\}, k \in \mathbb{Z}\}$ be the index set of the full univariate quarklet system and let g be the spline defined in Theorem 7.12. For $N \sim J^3$, $J \in \mathbb{N}$ there exist $c_\lambda \in \mathbb{R}$ such that*

$$g(x) = \sum_{\lambda \in \nabla' \subset \nabla^2 : |\nabla'| \leq N} c_\lambda \psi_\lambda(x), \quad x \in I, \tag{7.3.8}$$

$$|u_\alpha - g|^2_{H^1(I^2)} \lesssim \min\left(2^2, 2^{2\alpha-1}\right)^{-N^{1/3}} = e^{-\min(2, 2\alpha-1)\ln(2)N^{1/3}}. \tag{7.3.9}$$

Proof. Again, we derive a decomposition of the polynomials g_i with respect to elements of the tensor quarklet frame. From the tensor product structure

$$\psi_\lambda(x) = \psi_{p_1,j_1,k_1}(x_1)\psi_{p_2,j_2,k_2}(x_2),$$

the partition of unity

$$1 = \sum_{-m<k<m} \varphi_0(x_2 - k), \quad x_2 \in I,$$

and $g_i(x) = \tilde{g}_i(x_1)\chi_{[0,1]}(x_2)$ we conclude that only those frame elements of the form

$$\psi_\lambda(x) = \psi_{p_1,j_1,k_1}(x_1)\varphi_0(x_2 - k_2),$$

are needed for the decomposition of g_i. In particular we have the same decomposition as in the univariate case, therefore, employing the proof technique of Theorem 7.9, the asymptotic number of degrees of freedom follows. $\qquad\square$

Chapter 8

Numerical Experiments

This chapter is dedicated to numerical experiments. We perform different tests to support our theoretical findings. Firstly, in Section 8.1 we consider the numerical solution of elliptic PDEs on two-dimensional domains. In Section 8.2 we compare the performance of compression strategies based upon classical and second compression. In Section 8.3 we consider direct approximations of a certain univariate singularity function in terms of quarklets. The experiments have been performed with the software [48, 69].

8.1 The Poisson Equation on Domains

For our first numerical experiments we consider the Poisson equation with homogeneous Dirichlet boundary conditions on two-dimensional domains that can be decomposed into unit cubes as described in Section 5.2. Usually, those domains have reentrant corners that induce certain singular solutions to the Poisson equation, cf. [51]. It is well known that those solutions have a limited Sobolev regularity but arbitrary high Besov regularity. Therefore we expect a good performance of our adaptive schemes. Our test problem is given by the weak formulation

$$a(u, v) = f(v) \quad \text{for all } v \in H_0^1(\Omega), \tag{8.1.1}$$

where the bilinear form $a : H_0^1(\Omega) \times H_0^1(\Omega) \mapsto \mathbb{R}$ is defined by

$$a(u, v) = \sum_{k=1}^{2} \int_{\Omega} \frac{\partial u}{\partial x_k} \frac{\partial v}{\partial x_k} \, dx. \tag{8.1.2}$$

The L-shaped Domain

Let us mention that our first test case has already been considered in [27]. We begin with the most prominent example for the domains described above, namely the L-shaped domain domain $\Omega = (-1, 1)^2 \backslash [0, 1)^2$. To obtain a quarklet frame for Ω we split the domain as explained in Section 5.2, into the subdomains $\Omega_0^{(0)} = \{(-1, 0)\} + (0, 1)^2$, $\Omega_1^{(0)} = \{(-1, -1)\} + (0, 1)^2$ and $\Omega_2^{(0)} = \{(0, -1)\} + (0, 1)^2$. These subdomains with their incorporated boundary conditions are depicted in Figure 8.1. The arrows indicate the direction of the non-trivial extension. In our numerical experiments, we worked with a simple reflection operator. This operator is bounded on the energy space H^1, it is easy to implement and it produces favourable constants. By proceeding this way, conditions $(\mathcal{D}_1)$-$(\mathcal{D}_5)$ from Section 5.2 are fulfilled.

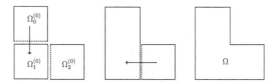

Figure 8.1: Extension process for the L-shaped domain. Dotted lines indicate free boundary conditions, straight lines indicate zero boundary conditions.

We denote the univariate quarklet frames as in (4.7.1). Then, the tensor product frames for the sub-cubes are

$$\Omega_0^{(0)}: \quad \Psi_0^1 = \Psi_{(1,1)}^1(\cdot + 1) \times \Psi_{(0,1)}^1,$$
$$\Omega_1^{(0)}: \quad \Psi_1^1 = \Psi_{(1,1)}^1(\cdot + 1) \times \Psi_{(1,1)}^1(\cdot + 1),$$
$$\Omega_2^{(0)}: \quad \Psi_2^1 = \Psi_{(0,1)}^1 \times \Psi_{(1,1)}^1(\cdot + 1).$$

To obtain a quarklet frame for $H_0^1(\Omega)$ we extend Ψ_0^1 and Ψ_2^1 as described in Chapter 5. Essentially this corresponds to reflecting those quarklets that do not vanish at the boundaries at the dotted lines in Figure 8.1. After that, we take the union of the two resulting sets of functions with Ψ_1^1. For the one-dimensional reference frame Ψ_σ^1 in $(0,1)$ we choose the biorthogonal spline wavelets of order $m = 3$ and $\tilde{m} = 3$ vanishing moments. We choose the right-hand side in (8.1.1) in such a way that the exact solution is the sum of $\sin(2\pi x)\sin(2\pi y)$, $(x,y) \in \Omega$ and the singularity function

$$\mathcal{S}(r,\theta) := 5\zeta(r)r^{2/3}\sin\left(\frac{2}{3}\theta\right), \qquad (8.1.3)$$

with (r,θ) denoting polar coordinates with respect to the re-entrant corner at the origin, and where ζ is a smooth truncation function on $[0,1]$, which is identically 1 on $[0,r_0]$ and 0 on $[r_1,1]$, for some $0 < r_0 < r_1 < 1$, see again [51] for details. Singularity functions of the form (8.1.3) are typical examples of functions that have a very high Besov regularity but a very limited L_2-Sobolev smoothness due to the strong gradient at the reentrant corner. Therefore, for this kind of solution it can be expected that adaptive (h-)algorithms outperform classical uniform schemes. We refer, e.g., to [22, 23] for a detailed discussion of these relationships.

We also expect that the very smooth sinusoidal part of the solution can be very well approximated by piecewise polynomials of high order. Therefore, our test example is contained in the class of problems for which we expect a strong performance of adaptive quarklet schemes.

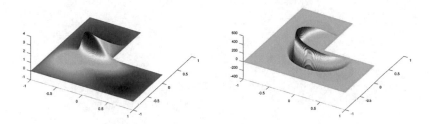

Figure 8.2: Exact solution and right-hand-side.

Figure 8.3 shows the adaptive refinement.

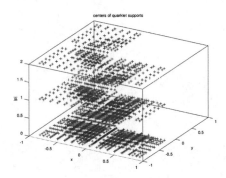

Figure 8.3: Center of quarklet supports.

To solve the problem numerically we utilise an adaptive version of the damped Richardson iteration as described in Section 6.1. In Figure 8.4 one can observe the ℓ_2-norm of the residual $\mathbf{Au}^{(j)} - \mathbf{f}$ plotted against the degrees of freedom of the approximants $\mathbf{u}^{(j)}$ and against the spent CPU time. We see that the algorithm is convergent with convergence order $\mathcal{O}(N^{-2})$. In [23] an adaptive wavelet frame approach based on overlapping domain decompositions was used to solve a similar problem. Since the singularity function (8.1.3) has arbitrary high Besov regularity, the convergence order of adaptive wavelet schemes only depend on the order of the underlying spline system. For $m = 3$, one gets the approximation rate $\mathcal{O}(N^{-1})$, see again [23, Sub. 6.2] for details. If we compare this to our approach we see that the adaptive quarklet schemes outperform the adaptive wavelet schemes in terms of degrees of freedom.

145

Our findings from Chapter 7 indicate that the observed rate of convergence is not optimal. We would conjecture that by employing a more sophisticated refinement strategy the order could be improved.

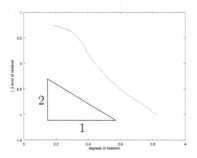 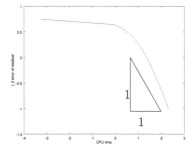

Figure 8.4: Adaptive error asymptotics.

The +-shaped Domain

The next test case is a +-shaped domain as depicted in Figure 8.5. It can be written as $\Omega = (-2,1)^2 \setminus \{[0,1)^2 \cup [0,1) \times (-2,1] \cup (-2,1]^2 \cup (-2,1] \times [0,1)\}$. We split this domain into the sub-cubes $\Omega_0^{(0)} = \{(-1,0)\} + (0,1)^2$, $\Omega_1^{(0)} = \{(-1,-1)\} + (0,1)^2$, $\Omega_2^{(0)} = \{(0,-1)\} + (0,1)^2$, $\Omega_3^{(0)} = \{(-1,-2)\} + (0,1)^2$ and $\Omega_4^{(0)} = \{(-2,-1)\} + (0,1)^2$. We equip the sub-cubes with the quarklet frames

$$\Omega_0^{(0)}: \quad \Psi_0^1 = \Psi_{(1,1)}^1(\cdot + 1) \times \Psi_{(0,1)}^1,$$

$$\Omega_1^{(0)}: \quad \Psi_1^1 = \Psi_{(1,1)}^1(\cdot + 1) \times \Psi_{(1,1)}^1(\cdot + 1),$$

$$\Omega_2^{(0)}: \quad \Psi_2^1 = \Psi_{(0,1)}^1 \times \Psi_{(1,1)}^1(\cdot + 1),$$

$$\Omega_3^{(0)}: \quad \Psi_3^1 = \Psi_{(1,1)}^1(\cdot + 1) \times \Psi_{(1,0)}^1(\cdot + 2),$$

$$\Omega_4^{(0)}: \quad \Psi_4^1 = \Psi_{(1,0)}^1(\cdot + 2) \times \Psi_{(1,1)}^1(\cdot + 1).$$

In contrast to the L-shaped domain, in this case extensions in all cardinal directions occur. Again, we choose the exact solution as the sum of $\sin(2\pi x)\sin(2\pi y)$ and the singularity function $\mathcal{S}$ given in (8.1.3). The following Figures show the solution, the adaptive refinement and the error asymptotics.

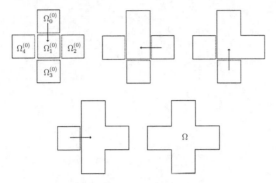

Figure 8.5: Extension process for the +-shaped domain. Dotted lines indicate free boundary conditions, straight lines indicate zero boundary conditions.

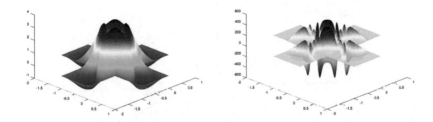

Figure 8.6: Exact solution and right-hand-side.

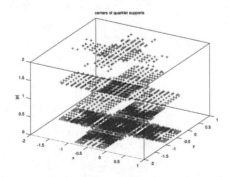

Figure 8.7: Center of quarklet supports.

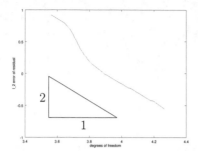 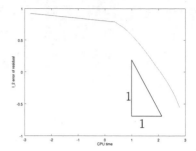

Figure 8.8: Adaptive error asymptotics.

The Snake-shaped Domain

Finally we consider the Snake-shaped domain $\Omega = (-1,2)^2 \setminus \{(-1,1] \times [1,2) \cup [0,2) \times (-1,0]\}$ as depicted in Figure 5.1. We split this domain into the sub-cubes $\Omega_0^{(0)} = \{(-1,-1)\} + (0,1)^2$, $\Omega_1^{(0)} = \{(-1,0)\} + (0,1)^2$, $\Omega_2^{(0)} = (0,1)^2$, $\Omega_3^{(0)} = \{(1,0)\} + (0,1)^2$ and $\Omega_4^{(0)} = \{(1,1)\} + (0,1)^2$. We equip the sub-cubes with the quarklet frames

$$\Omega_0^{(0)}: \quad \Psi_0^1 = \Psi_{(1,1)}^1(\cdot + 1) \times \Psi_{(0,1)}^1(\cdot + 1),$$

$$\Omega_1^{(0)}: \quad \Psi_1^1 = \Psi_{(1,1)}^1(\cdot + 1) \times \Psi_{(1,1)}^1,$$

$$\Omega_2^{(0)}: \quad \Psi_2^1 = \Psi_{(0,1)}^1 \times \Psi_{(1,1)}^1,$$

$$\Omega_3^{(0)}: \quad \Psi_3^1 = \Psi_{(1,1)}^1(\cdot - 1) \times \Psi_{(1,0)}^1,$$

$$\Omega_4^{(0)}: \quad \Psi_4^1 = \Psi_{(1,0)}^1(\cdot - 1) \times \Psi_{(1,1)}^1(\cdot - 1).$$

In contrast to the previous cases, we now have extensions from one sub-cube to more than one other sub-cube. Again, we choose the exact solution as the sum of $\sin(2\pi x)\sin(2\pi y)$ and the singularity function $\mathcal{S}$ given in (8.1.3). The following Figures show the solution, the adaptive refinement and the error asymptotics.

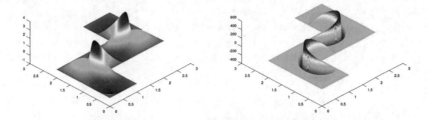

Figure 8.9: Exact solution and right-hand-side.

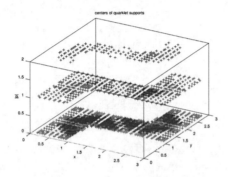

Figure 8.10: Center of quarklet supports.

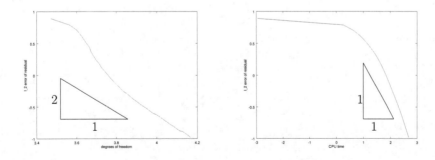

Figure 8.11: Adaptive error asymptotics.

Let us summarise the observations made in this section. We have seen that, for all our test cases, the adaptive algorithm performs consistently. The observed rate of convergence with respect to the degrees of freedom of $s = 2$ is superior compared to classical wavelet schemes. Our approximation results from Chapter 7 indicate that the convergence rate rate can be improved by employing an involved compression strategy. Furthermore, our approach of decomposing domains is not restricted to the L-shaped domain, but works also for more complicated domains. For tests on a non-Lipschitz domain we refer to [57].

8.2 Second Compression

In this section we consider the effect of second compression in the treatment of PDEs. We compare classical compression techniques and second compression techniques for a univariate test problem. Employing the cut-off rules from Theorems 6.4 and 6.7, respectively, lead to different performance of the implemented **APPLY** routines. In particular, second compression saves memory usage and computation time. Table 8.1 shows a comparison of **APPLY** routines. In our test case the saving was approximately 10 %, this effect will grow with increasing maximal levels j. Figure 8.12 shows the error $\|A - A_J\|_{\mathcal{L}(\ell_2)}$ of approximations to the stiffness matrix. Both classical and second compression give a better approximation than predicted. Furthermore, up to $J = 15$, the error of second compression is not significantly higher than classical compression. For $J > 15$ the strong decay of the classical compression error is a consequence of the implementation with a finite maximal level and polynomial degree. Finally, Figure 8.13 shows the error decay of an adaptive quarklet schemes using classical and second compression. It can bee seen that in particular in the initial phase of the algorithms second compression pays off. As a consequence of caching strategies in the implementation, the version using classical compression catches up.

Table 8.1: Numerical comparison of compression techniques.

	Second compression	Classical compression
Calls of $a(\cdot, \cdot)$ in one **APPLY** routine	376 626	421 916
Average time for one **APPLY** routine	3.9 sec	4.4 sec

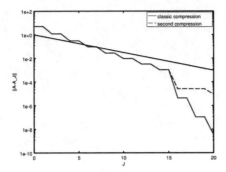

Figure 8.12: Blue: Classical numerical realisation of $\|A - A_J\|$,
Red: Numerical realisation of $\|A - A_J\|$ using second compression with
$J_1 = J, J_2 = J/2$ and $t = 1$,
Black: $2^{-J/2}$,
Parameters $m = \tilde{m} = 3, a = b = 2, \delta = 6, j_{max} = 11, p_{max} = 5$.

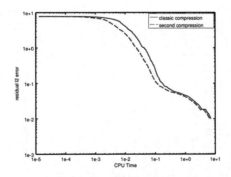

Figure 8.13: Blue: Classical compression in adaptive scheme,
Red: Second compression in adaptive scheme,
Parameters $m = \tilde{m} = 3, a = b = 2, \delta = 6, j_{max} = 8, p_{max} = 3$.

8.3 Adaptive Quarklet Approximation

In this section we conduct the direct approximation to the function u_α as described in Chapter 7. The geometric partition of the interval and the distribution of the polynomial degrees are a-priori known. Then, a suitable amount of fine quarks according to Theorem 7.5 is picked and, by using the developed reconstruction sequences, transformed into frame elements. The asymptotic rates are stated in Theorem 7.6 and 7.9. The following figures show the L_2 approximation error according to the number of degrees of freedom. Altogether the experiments confirm our theoretical findings. For the case $m = 2$, the experimental error decays even faster than the improved asymptotic rate of $e^{-\sqrt[3]{N}}$.

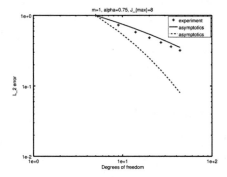

Figure 8.14: Error decay of quarklet approximation for $m = 1$, $J_{max} = 8$, $\alpha = 0.75$.

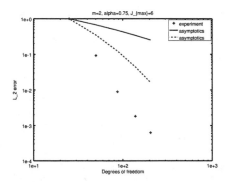

Figure 8.15: Error decay of quarklet approximation for $m = 2$, $J_{max} = 6$, $\alpha = 0.75$.

Discussion

In this thesis we have discussed several aspects concerning adaptive quarklet schemes. The initial question if it is possible to design hp versions of wavelet schemes clearly can be affirmed. The univariate findings from [28], in particular the frame property of quarklet systems and the compressibility of the resulting stiffness matrix, could be generalised to a class of domains in higher dimensions.

Let us go into details. In the shift-invariant setting, the construction of quarklets by keeping the underlying wavelet mask could be justified in hindsight. We showed that this construction provides a generalised multiresolution structure and allows for reconstruction properties. These findings also could be employed in the study of quarklet approximation. The characterisation of Besov spaces in the spirit of [30] could be transferred to the case of quarklet systems. The proof of the frame property for Sobolev spaces on domains was crucial for the subsequent design of adaptive quarklet schemes. Classical and second compression techniques have successfully been applied to the resulting stiffness matrix in the case of the Laplace operator. These findings in principle can be transferred to a wider class of operators. Finally, the exponential convergence of adaptive quarklet approximation gives strong hope that quarklet schemes converge fast. Our theoretical rate serves as a benchmark that solution methods for operator equations should achieve. It still is a long way to go to prove such convergence rates, but we can say that we essentially extended the theoretical foundation for adaptive quarklet schemes. The necessary preparations for our numerical experiments included the implementation of quarklet approximation, reconstruction methods, quarklets on the unit cube, suitable extension operators and various compression methods. Subsequent to this, our experiments affirmed our theoretical results.

Let us compare our findings with related work. Since hp versions of wavelet methods are quite innovative, it remains to discuss similarities and differences compared with hp-FEM. The optimal approximation rate in one dimension, cf. [2–5], was $e^{-N^{1/2}}$. Our approach delivers the rate $e^{-N^{1/3}}$. We conjecture that the weaker decay is an inevitable consequence of the multiscale structure. The asymptotic rates in two and three dimensions in [2] are $e^{-N^{1/3}}$ and $e^{-N^{1/5}}$, respectively. Since our systems contains anisotropic tensor products of univariate quarklets, our approximation rate does not depend on the dimension for certain singularities. For hp-FEM this exponential error decay has been observed in numerical experiments, see also [9, 47]. An important step in the development of adaptive quarklet schemes clearly is to design cut-off rules that make the exponential decay observable. Then, one of course has to drop the a-priori fixed distribution of levels and degrees. A promising approach seems to estimate the local regularity of the unknown solution and then to selectively increase either the level or the polynomial degree. For the case of hp-FEM, this has been done in [55]. To transfer this technique to the case of quarklets seems to be the next step in the development of adaptive quarklet schemes.

Let us mention two further aspects in the context of adaptive quarklet schemes that were not discussed in full length in the course of this thesis.

Orthogonal Polynomials

The original construction in [68] was based upon orthogonal polynomials, furthermore in [57] it was claimed that enrichment with orthogonal polynomials was a promising approach. This approach clearly has the advantage that not only the quarklets, but the quark generators have vanishing moments. This can lead to better compression results. Let us first discuss the construction. In [56] it was shown that in the shift-invariant case it is possible to construct quarklets with increasing vanishing moments. Let φ be the cardinal B-spline of fixed order m. Let $\{\gamma_p\}_{p \geq 0}$ denote the φ-orthogonal polynomials, that means

$$\langle \gamma_q, \gamma_p \rangle_\varphi := \int_{\mathbb{R}} \gamma_p(x) \gamma_q(x) \varphi(x) \, \mathrm{d}x = 0, \quad 0 \leq q < p.$$

Now the quark φ_p can be defined by

$$\varphi_p := \gamma_p \varphi.$$

It immediately follows that the quark φ_p itself has p vanishing moments. By arguments similar as in [28, Lem. 2] one can show that the quarklet defined by (4.2.1) has $\tilde{m} + p$ vanishing moments. Moreover, this construction can be adapted to the interval. Let again $\varphi_{j,k}$ be the generators of the Primbs basis. Let further $\{\gamma_{p,j,k}\}_{p \geq 0}$ denote orthogonal polynomials with respect to the generator $\varphi_{j,k}$. Note that for the inner generators these polynomials coincide with γ_p. Analogously to the construction presented in Section 4.3 we look for solutions $b_{k,l}^{p,j,\vec{\sigma}}$ for the system of equations

$$\sum_{l=-m+1+\mathrm{sgn}\,\sigma_l+k}^{-m+1+\mathrm{sgn}\,\sigma_l+k+\tilde{m}} b_{k,l}^{p,j,\vec{\sigma}} \int_{\mathbb{R}} x^q \gamma_{p,j,l}(x) \varphi_{j+1,l}(x) \, \mathrm{d}x = 0, \quad q = p, ..., \tilde{m} + p - 1.$$

By orthogonality we already have

$$\int_{\mathbb{R}} x^q \gamma_{p,j,l}(x) \varphi_{j+1,l}(x) \, \mathrm{d}x = 0, \quad q = 0, \ldots, p - 1,$$

hence the quarklet defined by (4.3.1) has $\tilde{m} + p$ vanishing moments. Now we turn over to compression properties. To prove the cancellation property (6.2.12), we used the full amount of vanishing moments. This is only possible for quarklets with non-intersecting singular supports. In general, one is limited by the quarklets' global regularity, see (6.2.4). Accordingly, exploiting the increasing vanishing moments is only possible with second compression techniques. Let ψ_λ' be a quarklet with $\tilde{m} + p$ vanishing moments. Let $\lambda \in \bigcup_{j,i} A_{j,\lambda',i}$. Then we can deduce the following cancellation property for unweighted quarklets:

$$|\langle \psi_\lambda', \psi_{\lambda'}' \rangle_{L_2(I)}| \lesssim 2^{(j+j')} 2^{-(j-j')(\tilde{m}+p+\frac{1}{2})} (p'+1)^{2\tilde{m}-m+2p+3} (p+1)^{-(m-2)}.$$

To dampen the exponential growth in this term one would need exponentially decreasing weights with respect to the polynomial degree. On the other hand, for exponential weights in adaptive schemes we expect a strong repression of the ansatz functions with a higher polynomial degree. Therefore we think that the orthogonal polynomial approach is not favourable.

Sparsity

The redundancy of our quarklet frames leads to the question, if any numerical scheme automatically chooses the most appropriate frame elements to obtain a *sparse* decomposition of a certain function. It is possible to enforce sparsity in adaptive schemes such as the iteration (6.1.10), cf. [29, 42]. One looks for quarklet coefficients which fulfil

$$\min_c \|\boldsymbol{A}\boldsymbol{c} - \boldsymbol{f}\|_{\ell_2(I)} + \mu \|\boldsymbol{c}\|_{\ell_1(I)}, \quad \mu \in \mathbb{R}.$$

This leads to the normal equation

$$\boldsymbol{A}^T \boldsymbol{A} \boldsymbol{c} = \boldsymbol{A}^T \boldsymbol{f} - \frac{1}{2}\vec{\mu},$$

which can iteratively be solved. With a suitable shrinkage operator S_μ, the *Landweber iteration* is given by

$$\boldsymbol{c}^{(k+1)} = S_\mu \left(\boldsymbol{c}^{(k)} + \omega(\boldsymbol{A}^T \boldsymbol{f} - \frac{1}{2}\vec{\mu} - \boldsymbol{A}^T \boldsymbol{A} \boldsymbol{c}^{(k)}) \right).$$

One obvious drawback is the squaring of the condition number, furthermore it is not guaranteed that the resulting sparse representation is appropriate in the sense of Chapter 7, where a certain geometric partition of the interval and a certain degree distribution was used.

List of Figures

List of Tables

List of Symbols

$\Delta_{j,\vec{\sigma}}$	Boundary adapted translation parameter set 36
Δ_j	Boundary adapted translation parameter set 34
$\langle\cdot,\cdot\rangle_{H^m(D)}$	H^m scalar product 11
$\langle\cdot,\cdot\rangle_{L_2(D)}$	L_2 scalar product 10
Δ	Laplace operator 15
∇_σ	Multivariate quarklet index set 86
$\nabla_{j,\sigma}$	Multivariate translation parameter set 112
∇	Quarklet index set on domain 96
Δ_h^m	Discrete difference operator for measurable functions 12
$\lfloor s \rfloor$	Largest integer not bigger than s 11
$[\cdot,\cdot](z)$	26
$\nabla_{\vec{\sigma}}^R$	Boundary adapted wavelet basis index set 36
$\nabla_{\vec{\sigma}}$	Boundary adapted quarklet index set 71
$\nabla_{j,\vec{\sigma}}$	Boundary adapted translation parameter set 67
$(x)_+$	Truncated monomial $(x)_+ := \max\{x,0\}$ 30
$\mathscr{A}(z)$	Symbol matrix of a refinable function vector 40
$A_{j,\lambda',i}$	103
$\mathscr{A}_\rho(z^2)$	Subsymbol matrix 40
$\mathbf{A}_J$	Truncated stiffness matrix 102, 105, 114, 119, 123
$\mathbf{A}$	Biinfinite stiffness matrix 100, 102
$a(z)$	Symbol of a refinable function 25
$a_{\lambda,\lambda'}$	Stiffness matrix entry 113
$a_\rho(z^2)$	Subsymbol 29
$a(\cdot,\cdot)$	Bilinear form 16
$\mathscr{B}(z)$	Symbol matrix of a wavelet vector 41
$B_1(0)$	Unit ball in the euclidean plane 12
$B_{j,k}^m$	Schoenberg B-spline 35
$b(z)$	Symbol of a wavelet 27
$b_\rho(z^2)$	Subsymbol 29
$B_q^s(L_p(D))$	Besov space with positive smoothness 13
$c_\rho(z^2)$	Subsymbol 29

Bibliography

[1] R. A. Adams, *Sobolev spaces*, Academic Press, New York, 1975.

[2] I. Babuška and M. Suri, *The p- and h-p versions of the finite element method, basic principles and properties*, SIAM Review **36** (1994), no. 4, 578–632.

[3] I. Babuška and W. Gui, *The h, p and h-p versions of the finite element method in 1 dimension. Part II. The error analysis of the h- and h-p versions.*, Numerische Mathematik **49** (1986), no. 4, 613–658.

[4] _____, *The h, p and h-p versions of the finite element method in 1 dimension. part III. The adaptive h-p version.*, Numerische Mathematik **49** (1986), no. 4, 659–683.

[5] I. Babuška and W. Gui, *The h, p and h-p versions of the finite element method in 1 dimension. I. The error analysis of the p-version.*, Numer. Math. **49** (1986), 577–612 (English).

[6] I. Babuška and B. Guo, *The h-p version of the finite element method. I. The basic approximation results.*, Comput. Mech. **1** (1986), 21–41 (English).

[7] _____, *The h-p version of the finite element method. II. General results and applications.*, Comput. Mech. **1** (1986), 203–220 (English).

[8] P. Binev, W. Dahmen, and R. DeVore, *Adaptive finite element methods with convergence rates*, Numer. Math. **97** (2004), no. 2, 219–268.

[9] M. Bürg and W. Dörfler, *Convergence of an adaptive hp finite element strategy in higher space-dimensions.*, Appl. Numer. Math. **61** (2011), no. 11, 1132–1146 (English).

[10] C. Canuto, R.H. Nochetto, R. Stevenson, and M. Verani, *High-order adaptive Galerkin methods*, Spectral and High Order Methods for Partial Differential Equations ICOSAHOM 2014 (M. Berzins, J.S. Hesthaven, and R.M. Kirby, eds.), Lect. Notes Comput. Sci. Eng., no. 106, Springer, 2014, pp. 51–72.

[11] _____, *Convergence and optimality of hp-AFEM*, Numer. Math. **135** (2017), no. 4, 1073–1119.

[12] _____, *On p-robust saturation for hp-AFEM*, Comput. Math. Appl. **73** (2017), no. 9, 2004–2022.

[13] N. Chegini, S. Dahlke, U. Friedrich, and R. Stevenson, *Piecewise tensor product wavelet bases by extensions and approximation rates*, Found. Comput. Math. **82** (2013), 2157–2190.

[14] S.S. Chen, D.L. Donoho, and M.A. Saunders, *Atomic decomposition by basis pursuit*, SIAM Rev. **43** (2001), no. 1, 129–159.

[15] O. Christensen, *Frames and Bases, an Inroductory Course*, Birkhäuser, Basel, 2008.

[16] C. K. Chui, *An Introduction to Wavelets*, Academic Press, Boston, 1992.

[17] P. G. Ciarlet, *The Finite Element Method for Elliptic Problems*, SIAM, 2002.

[18] A. Cohen, *Numerical Analysis of Wavelet Methods*, Studies in Mathematics and its Applications, vol. 32, North-Holland, Amsterdam, 2003.

[19] A. Cohen, W. Dahmen, and R. DeVore, *Adaptive wavelet methods for elliptic operator equations – Convergence rates*, Math. Comput. **70** (2001), no. 233, 27–75.

[20] ――――, *Adaptive wavelet methods II: Beyond the elliptic case*, Found. Comput. Math. **2** (2002), no. 3, 203–245.

[21] A. Cohen, I. Daubechies, and J.-C. Feauveau, *Biorthogonal bases of compactly supported wavelets*, Commun. Pure Appl. Math. **45** (1992), 485–560.

[22] S. Dahlke, W. Dahmen, and R. DeVore, *Nonlinear approximation and adaptive techniques for solving elliptic operator equations*, Multiscale Wavelet Methods for Partial Differential Equations (W. Dahmen, A. Kurdila, and P. Oswald, eds.), Academic Press, San Diego, 1997, pp. 237–283.

[23] S. Dahlke, M. Fornasier, M. Primbs, T. Raasch, and M. Werner, *Nonlinear and adaptive frame approximation schemes for elliptic PDEs: Theory and numerical experiments*, Numer. Methods Partial Differ. Equations **25** (2009), no. 6, 1366–1401.

[24] S. Dahlke, M. Fornasier, and T. Raasch, *Adaptive frame methods for elliptic operator equations*, Adv. Comput. Math. **27** (2007), no. 1, 27–63.

[25] S. Dahlke, M. Fornasier, T. Raasch, R. Stevenson, and M. Werner, *Adaptive frame methods for elliptic operator equations: The steepest descent approach*, IMA J. Numer. Anal. **27** (2007), no. 4, 717–740.

[26] S. Dahlke, U. Friedrich, P. Keding, T. Raasch, and A. Sieber, *Adaptive quarkonial domain decomposition methods for elliptic partial differential equations*, Bericht 2018–1, Philipps Universität Marburg, 2018.

[27] _____, *Adaptive quarkonial domain decomposition methods for elliptic partial differential equations*, IMA J. Numer. Anal. (2020), to appear.

[28] S. Dahlke, P. Keding, and T. Raasch, *Quarkonial frames with compression properties*, Calcolo **54** (2017), no. 3, 823–855.

[29] S. Dahlke, G. Kutyniok, G. Steidl, and G. Teschke, *Shearlet coorbit spaces and associated Banach frames.*, Appl. Comput. Harmon. Anal. **27** (2009), no. 2, 195–214.

[30] S. Dahlke, P. Oswald, and T. Raasch, *A note on quarkonial systems and multilevel partition of unity methods*, Mathematische Nachrichten **286** (2013), 600–613.

[31] S. Dahlke, T. Raasch, and A. Sieber, *Exponential convergence of adaptive quarklet approximation*, J. Complexity (2020), to appear.

[32] W. Dahmen, *Wavelet and multiscale methods for operator equations*, Acta Numerica **6** (1997), 55–228.

[33] W. Dahmen, B. Han, R. Q. Jia, and A. Kunoth, *Biorthogonal multiwavelets on the interval: Cubic Hermite splines*, Constr. Approx. **16** (2000), no. 2, 221–259.

[34] W. Dahmen, H. Harbrecht, and R. Schneider, *Compression techniques for boundary integral equations – asymptotically optimal complexity estimates*, SIAM J. Numer. Anal. **43** (2006), no. 6, 2251–2271.

[35] W. Dahmen, A. Kunoth, and K. Urban, *A wavelet-Galerkin method for the Stokes problem*, Computing **56** (1996), 259–302.

[36] _____, *Biorthogonal spline-wavelets on the interval — Stability and moment conditions*, Appl. Comput. Harmon. Anal. **6** (1999), 132–196.

[37] W. Dahmen and R. Schneider, *Wavelets with complementary boundary conditions — Function spaces on the cube*, Result. Math. **34** (1998), no. 3–4, 255–293.

[38] _____, *Composite wavelet bases for operator equations*, Math. Comput. **68** (1999), 1533–1567.

[39] _____, *Wavelets on manifolds I. Construction and domain decomposition*, SIAM J. Math. Anal. **31** (1999), 184–230.

[40] I. Daubechies, *Orthonormal bases of compactly supported wavelets*, Commun. Pure Appl. Math. **41** (1988), no. 7, 909–996.

[41] _____, *Ten Lectures on Wavelets*, CBMS–NSF Regional Conference Series in Applied Math., vol. 61, SIAM, Philadelphia, 1992.

[42] I. Daubechies, M. Defrise, and C. De Mol, *An iterative thresholding algorithm for linear inverse problems with a sparsity constraint*, Commun. Pure Appl. Math. **57** (2004), no. 11, 1413–1457.

[43] R. DeVore and G.G. Lorentz, *Constructive Approximation*, Springer, Berlin, 1998.

[44] T. J. Dijkema, *Adaptive Tensor Product Wavelet Methods for Solving PDEs*, Ph.D. thesis, Utrecht University, 2009.

[45] M. Dobrowolski, *Angewandte Funktionalanalysis*, Springer, Berlin Heidelberg, 2006.

[46] D. L. Donoho, *Compressed sensing.*, IEEE Trans. Inf. Theory **52** (2006), no. 4, 1289–1306 (English).

[47] W. Dörfler and V. Heuveline, *Convergence of an adaptive hp finite element strategy in one space dimension.*, Appl. Numer. Math. **57** (2007), no. 10, 1108–1124 (English).

[48] J. W. Eaton, D. Bateman, S. Hauberg, and R. Wehbring, *GNU Octave version 4.0.0 manual: a high-level interactive language for numerical computations*, 2015.

[49] L.C. Evans, *Partial differential equations*, Graduate Studies in Mathematics, vol. 19, American Mathematical Society (AMS), Providence, RI, 1998.

[50] M. Griebel and P. Oswald, *Tensor product type subspace splittings and multilevel iterative methods for anisotropic problems*, ACMA **4** (1995), 171–206.

[51] P. Grisvard, *Elliptic Problems in Nonsmooth Domains*, Pitman Publishing, Boston-London-Melbourne, 1985.

[52] K.-H. Gröchenig, *Foundations of Time–Frequency Analysis*, Birkhäuser, Boston-Basel-Berlin, 2001.

[53] A. Haar, *Zur Theorie der orthogonalen Funktionensysteme*, Math. Annalen **69** (1910), 331–371.

[54] W. Hackbusch, *Elliptic Differential Equations. Theory and Numerical Treatment*, 2nd ed., Springer, Berlin, 2017.

[55] P. Houston, B. Senior, and E. Süli, *Sobolev regularity estimation for hp-adaptive finite element methods.*, Numerical mathematics and advanced applications. Proceedings of ENUMATH 2001, the 4th European conference, Ischia, July 2001, Berlin: Springer, 2003, pp. 631–656 (English).

[56] P. Keding, *Quarklets with increasing vanishing moments*, private communication, December 2017.

[57] _____, *Quarklets: Construction and Application in Adaptive Frame Methods*, Ph.D. thesis, Fachbereich Mathematik und Informatik, Philipps-Universität Marburg, 2018.

[58] A. Khosravi and M. S. Asgari, *Frames and bases in tensor product of hilbert spaces.*, Int. Math. J. **4** (2003), no. 6, 527–537.

[59] K. Koch, *Interpolating scaling vectors and multiwavelets in* $\mathbb{R}^d$, Ph.D. thesis, Fachbereich Mathematik und Informatik, Philipps-Universität Marburg, 2006.

[60] S. Mallat, *Multiresolution approximation and wavelet orthonormal bases of* $L_2(\mathbb{R}^d)$, Trans. Amer. Math. Soc. **315** (1989), 69–87.

[61] Y. Meyer, *Principe d'incertitude, bases hilbertiennes et algèbres d'opérateurs*, Sémin. Bourbaki **38** (1985/86), no. 662, 209–223.

[62] R.H. Nochetto, K. Siebert, and A. Veeser, *Theory of adaptive finite element methods: An introduction*, Multiscale, Nonlinear and Adaptive Approximation, Springer, Berlin, 2009, pp. 409–542.

[63] P. Oswald, *Note on 1d quarklet approximation*, private communication, September 2019.

[64] G. Plonka, *Approximation order provided by refinable function vectors*, Constr. Approx. **13** (1997), 221–244.

[65] G. Plonka and V. Strela, *From wavelets to multiwavelets*, Mathematical Methods of Curves and Surfaces II (Nashville) (M. Daehlen, T. Lyche, and L. L. Schumaker, eds.), Vanderbilt University Press, 1998, pp. 1–25.

[66] M. Primbs, *Stabile biorthogonale Spline-Waveletbasen auf dem Intervall*, Ph.D. thesis, Universität Duisburg-Essen, 2006.

[67] M. Primbs, *New stable biorthogonal spline-wavelets on the interval*, Results in Mathematics **57** (2010), no. 1, 121–162.

[68] T. Raasch, *Quarkonial frames of wavelet type: Stability and moment conditions*, Talk at Dagstuhl Seminar 11051, February 2011.

[69] T. Raasch, M. Werner, U. Friedrich, S. Kinzel, C. Hartmann, P. Keding, and A. Sieber, *Wavelet and Multiscale Library*, `https://wavelet-and-multiscale-library.github.io/`, 2000–2019.

[70] E. A. Robinson, *Time series analysis and applications.*, Houston, Texas: Goose Pond Press. V, 621 p., 1981.

[71] W. Rudin, *Functional Analysis*, McGrawHill, 1991.

[72] Walter Rudin, *Real and complex analysis*, 3rd ed., New York, NY: McGraw-Hill, 1987 (English).

[73] C. Schneider, *Besov spaces with positive smoothness*, Ph.D. thesis, Universität Leipzig, 2009.

[74] R. Schneider, *Multiskalen- und Wavelet-Matrixkompression. Analysisbasierte Methoden zur effizienten Lösung großer vollbesetzter Gleichungssysteme*, Habilitationsschrift, TH Darmstadt, 1995.

[75] L. L. Schumaker, *Spline functions : Basic theory*, Cambridge University Press, 2007.

[76] C. Schwab, *p- and hp-Finite Element Methods. Theory and Applications in Solid and Fluid Mechanics*, Clarendon Press, Oxford, 1998.

[77] A. Sieber, *Adaptive Quarklet-Verfahren für Dirichlet-Randwertprobleme*, Master's thesis, Philipps-Universität Marburg, 2017.

[78] J. R. Silvester, *Determinants of block matrices*, The Mathematical Gazette **84** (2000), no. 501, 460–467.

[79] R. Stevenson, *Adaptive solution of operator equations using wavelet frames*, SIAM J. Numer. Anal. **41** (2003), no. 3, 1074–1100.

[80] ———, *On the compressibility of operators in wavelet coordinates*, SIAM J. Math. Anal. **35** (2004), no. 5, 1110–1132.

[81] ———, *Adaptive wavelet methods for solving operator equations: an overview.*, DeVore, Ronald (ed.) et al., Multiscale, nonlinear and adaptive approximation. Dedicated to Wolfgang Dahmen on the occasion of his 60th birthday. Springer, Berlin, 2009, pp. 543–597.

[82] The Sage Developers, *Sagemath, the Sage Mathematics Software System (Version 8.6)*, 2019, https://www.sagemath.org.

[83] A. F. Timan, *Theory of approximation of functions of a real variable. Translated by J.Berry.*, International Series of Monographs on Pure and Applied Mathematics. 34. Oxford etc.: Pergamon Press. XII, 631 p. (1963)., 1963.

[84] H. Triebel, *Interpolation Theory, Function Spaces, Differential Operators*, North-Holland, Amsterdam, 1978.

[85] ———, *Theory of Function Spaces*, Birkhäuser, Basel, 1983.

[86] ———, *Theory of Function Spaces II*, Birkhäuser, Basel, 1992.

[87] ———, *Fractals and Spectra related to Fourier Analysis and Function Spaces*, Birkhäuser, Basel, 1997.

[88] ———, *Theory of Function Spaces III*, Birkhäuser, Basel, 2006.

[89] R. Verfürth, *A Review of A Posteriori Eerror Estimation and Adaptive Mesh-Refinement Techniques*, Wiley-Teubner, Chichester, UK, 1996.

[90] D. Werner, *Funktionalanalysis*, 6. ed., Springer, Berlin, 2007.

[91] M. Werner, *Adaptive Wavelet Frame Domain Decomposition Methods for Elliptic Operators*, Ph.D. thesis, Fachbereich Mathematik und Informatik, Philipps-Universität Marburg, 2009.

[92] P. Wojtaszczyk, *A Mathematical Introduction to Wavelets*, Cambridge University Press, Cambridge, 1997.